Drainage Engineering

Related Books

Title	*Auhtor*
— Agricultural Drainage Engineering	*H.V. Parmar*
— Agriculture Chemical Users Manual	*Geoff Cowles*
— Basic Statistics *2nd Ed*	*P.K. Mohanty*
— Beverages: Processing and Technology	*Deepak Mudgil*
— Drainage Engineering: Principles and Practices	*S.K. Gupta*
— Drought Mitigation and Management	*Suresh Kumar*
— Drought Proofing and Water Resource Management	*T.S. Chouhan*
— Dryland Technology *2nd Ed*	*M.L. Jat*
— Functional Foods: Sources and Health Benefits	*Deepak Mudgil*
— Greenhouse Engineering	*R.A. Aldrich*
— Ground Water Manual a Water Resources Technical Publication	*USDI*
— Handbook on Pressurized Irrigation Techniques *2nd Ed*	*A. Phocaides*
— Horticultural Engineering Technology Field Machinery	*R.C. Balls*
— Irrigation Resources	*S.B. Verma*
— Management of Saline & Waste Water in Agriculture	*S.K. Gupta*
— Principles in the Quantitative Analysis of Water Fertilizers Plants and Soils	*U.S.S. Ramulu*
— Principles of Agricultural Meteorology	*B.M. Mote*
— Processing Tropical Crops: A Technological Approach	*J.J. Asiedu*
— Quinoa *(Chenopodium Quinoa Willd)*	*Dharm Singh*
— Rainfed Farming Development in Central India	*G.P. Verma*
— Rainwater Management Theory and Practice	*M.L. Jat*
— Remote Sensing and GIS Applications: A Starter Guide	*D.C. Joshi*
— Remote Sensing and GIS GPS Based Resource Management	*T.S. Chouhan*
— Remote Sensing for Natural Resources Management & Monitoring	*Mahesh Gaur*
— Remote Sensing: Principles and Applications *2nd Ed*	*A.N. Patel*
— Standard Methods for Analysis of Soil Plant and Water	*I.C. Gupta*
— Sustainable Development of Dryland Agriculture in India	*R.P. Singh*
— Sustainable Dryland Farming	*A.M. Dhopte*
— Tamarind Science and Technology	*A. Balasubramanian*
— Testing of Agriculture Technological Processes: A Systems Approach	*S.V. Kardashevskii*
— Theory, Construction and Calculation Agricultural Machines (Vol. 1)	*E.S. Boson*
— Theory, Construction and Calculation Agricultural Machines (Vol. 2)	*E.S. Boson*
— Use of Saline Water in Agriculture 3rd Revised Ed	*I.C. Gupta*
— Water Measurement Manual 3rd Ed	*USDI*

Drainage Engineering

James N. Luthin
Professor of Water Science and Civil Engineering
University of California,
Davis

United Book Prints

Reprinted in India by:

UNITED BOOK PRINTS
5-B, Bhagat ki Kothhi, Near Police Chowki
New Pali Road, Jodhpur 342 001 (INDIA)
E-mail: info@scientificpub.com

Print : 2021

ISBN: 978-81-92686-08-0
eISBN: 978-93-88148-17-7

Printed in India

PREFACE

This book is intended as a textbook in drainage engineering. It should be of use to those in agricultural and civil engineering. Parts of the book may be useful to soil physicists as well.

Drainage engineering is truly an interdisciplinary area of research, development, and practice. It is not possible, in a book of this size, to present completely the background in soils and hydraulics needed for drainage work. Some knowledge of soils is assumed as well as a knowledge of elementary fluid mechanics.

The last two decades have seen great advances in the art and science of drainage. Scientists and engineers all over the world have made many contributions to our knowledge of the movement of water through soils and its control. It is now possible to design drainage systems based upon measurable soil properties. Much, of course, remains to be done, especially in terms of the effect of water tables on plants.

It is the purpose of this book to present an approach to the problems of drainage based upon developments of the last several decades. Much of the theoretical work must necessarily be excluded from the book. I have tried to present the theories and practices which are in common use. This does not mean that more accurate theories do not exist. However, I have tried to limit the discussion to methods which are commonly used today.

All of us who work in the field of drainage owe a great debt to Dr. S. B. Hooghoudt, the first man to present a complete rational analysis of the drainage problem. His work remains today an outstanding and classical piece of research and development. Though his theories have been criticized because of some of the simplifying assumptions, they are still very close to much more exact and complicated developments. I would be remiss if I did not indicate also the contributions to drainage by Dr. E. C. Childs in England and by Dr. Don Kirkham in the United States. The stimulus which these two individuals have given to drainage research cannot be ignored.

Many people have given generously of their time and effort in the preparation of the book. I hesitate to list them for fear I might inadvertently leave a name out of the list. However, I do want to express my deep appreciation for the help and information my fellow workers in drainage have given me.

JAMES N. LUTHIN

Davis, California
October, 1965

CONTENTS

Chapter 1 INTRODUCTION

THE NEED FOR DRAINAGE

A drainage problem is caused by an excess of water either on the surface of the soil or in the root zone beneath the surface of the soil. If the water stands on the surface of the soil the problem is one of surface drainage. This more obvious type of drainage problem can be remedied by providing some method of removing the surface water.

The other type of drainage problem concerns water that occurs beneath the surface of the soil. A high water table is present. Frequently the presence of the high water table is not evident from an inspection of the soil surface. In many instances the soil surface may appear to be dry, although waterlogged soil at depths of two or more feet beneath the surface may cause serious damage to the crops which are being grown. Methods of investigating such problems form an important part of the engineering of a drainage system.

PURPOSE OF DRAINAGE

The main purpose of drainage is to provide a root environment that is suitable for the maximum growth of plants. The object of providing drainage is to increase production and to sustain yields over long periods of time. One of the main reasons that poor drainage causes a decrease in crop production is the fact that the plant roots have only a limited amount of soil in which to grow. This means that the plant root system is not adequate to supply the top of the plant with the foods it needs. Plants do not do well under such circumstances. Not only do they lack food, but because of the inadequate root system they may even suffer from lack of water when the water table drops. An additional factor which influences plant growth in poorly drained soils is the lack of certain plant nutrients under waterlogged conditions.

EFFECT OF POOR DRAINAGE ON SOIL AND PLANTS

Water that fills the soil pores not only displaces the air in the soil but also obstructs the gases which are given off by the roots. These soil channels need be blocked only at one point to become quite inefficient for the exchange between the gases in the soil and those in the outside atmosphere. The oxygen content in wet soils is limited not only because of the small amount of oxygen dissolved in the water, but also because of the extremely slow rate of diffusion of gases through such soils.

In waterlogged soils gas exchange is confined to a fraction of the top inch of soil. Below this, free oxygen is virtually nonexistent. When soils are flooded the oxygen in the soil and in the water disappears within a few hours.

Figure 1-1 A wet spot with cowslips and willows growing in and around it. Crops cannot be grown here because of the high water-table conditions. Photo courtesy Department of Agricultural Engineering, Cornell University.

Even if oxygen is then applied by artificial aeration the newly applied oxygen disappears as rapidly.

After the dissolved oxygen in waterlogged soils has been consumed, anaerobic decomposition of organic matter takes place. This results in the production of reduced organic compounds such as methane or marsh gas, methyl compounds, and complex aldehydes. Mineral substances in the soil are altered from the oxidized state to a reduced state. It has been found that toxic concentrations of ferrous and sulfide ions may develop within a few days after submergence of the soil. Toxic concentrations of manganous ions take somewhat longer to develop.

Waterlogging generally leads to a deceleration in the rate of the decomposition of organic matter, which means that organic matter accumulates after waterlogging. Because of the slowing down in the rate of decomposition, nitrogen tends to remain locked up in the organic residues. Nitrogen, therefore, is often a limiting factor to plant growth on poorly drained soils.

In drained soil the mineralization of nitrogen, that is, the release of nitrogen upon the decomposition of organic matter, proceeds at a steady rate. In waterlogged soils, however, the rate of mineralization decreases rapidly after an initial period of rapid release.

Figure 1-2 The same field shown in Figure 1-1 fourteen months later after a single line of drain tile has been installed.

The decline in the rate of transpiration that takes place when plants are flooded reflects the difficulty plants experience in taking up moisture from waterlogged soils. This inhibited uptake is related to deficient oxygen and increased concentrations of carbon dioxide under such conditions. These conditions persist due to the decay of roots and the lack of formation of new roots.

Waterlogging has an effect on the uptake of nutrients by plants. This is shown by certain symptoms which develop under circumstances of waterlogging. These symptoms may be yellowing, reddening, or a scorched or stippled appearance of the leaves. These symptoms, under other conditions, may indicate an unbalance in nutrient supply.

It was pointed out earlier that toxic concentrations of ferrous and sulfide ions show up soon after waterlogging, while organic compounds harmful to plant growth may be produced by anaerobic decomposition. One of these compounds is methane, which has been found to inhibit the growth of tomato plants completely, and which affects barley more adversely than does aeration by nitrogen gas.

Let us turn to another phase of water tables and their effects on plant growth, and consider those crop responses to high water tables which are not necessarily injurious. It has been found, in many parts of the world, that a plant can extract considerable amounts of the water from a high water table. In fact, even though the water table may be very deep, plants can still extract appreciable amounts of water. This has been shown in desert areas such as the Escalante Valley of Utah, where greasewood and similar species have been found to extract from water tables that are over twenty feet below the ground surface. Similarly, it has been found that crops such as alfalfa can extract large amounts of water from water tables that may be considerable depths below the soil surface. In The Netherlands the people rely on a high water table to supply a large amount of water for the plant growth. But in our discussion of high water tables and their beneficial effects on plant growth, we must continually bear in mind that there must be some control of the water table if any benefits are to be realized. A water table which fluctuates from the ground surface to some distance below the ground surface cannot be considered a benefit; it must be considered a liability. Fluctuating water tables cause the plant roots to rot off. When the water table drops there are no roots for the plants to use to extract moisture from the wet-root zone. The plant suffers from a lack of water and the yields are reduced accordingly.

We should note that not all soils are good for supplying water from water tables to plant roots. In some soils the rate of capillary rise is so slow that the plants do not get enough water from the water table. This would be particularly true in heavy, dense clay soils where the rate of rise of water from the water table would not be equal to the rate of transpiration. In such soils deep drainage would certainly be the recommended practice. In some of the sandy loam soils the rate of capillary rise may be very rapid and in these soils it may be possible for the water table to supply an appreciable amount of the water that the plant needs.

It is generally recognized that well-drained soils are required for fruit growing. Reports of disease conditions associated with excess moisture are numerous. For example, in Oregon it has been determined that orchards ordinarily require a depth of water table of six to eight feet. A higher water table persisting for three or four days following rain or irrigation during growing season does no harm.

In many cases the benefits obtained by removing excess water have been very conspicuous and have led to a hundredfold increase in land values. This applies in particular to swamp areas where agriculture was restricted or not possible at all, and where drainage has created valuable agricultural lands.

We can summarize our understanding of the effect on plants of poor drainage or high water tables by saying that, in general, dissolved oxygen is virtually absent in submerged soils, apparently because of microbiological activity. Reducing processes in the soil set in quite soon after waterlogging, and lead to rates of decomposition of organic matter which are lower than those in well-drained soils. An accumulation of reduced iron, manganese, and sulfur may develop soon after waterlogging, which, in combination with reduced or partly oxidized products of the decomposition of organic matter, may be harmful to crop production. Phosphorus becomes more soluble and tends to be leached out of the soil.

OTHER BENEFITS FROM DRAINAGE

The ponding of water during the summer may cause "scalding" of the crops. In areas having hot summer temperatures, water that is ponded

Figure 1-3 Scalding of alfalfa due to surface water standing for several days following irrigation.

because of irrigation or rainfall, will kill planted grasses or legumes. Drainage of the land by proper grading will reduce the damage due to scalding and will increase the yield.

A health hazard is created by mosquitoes which breed in ponds and small puddles in the field. If the fields are properly graded and proper outlets and ditches for the water are provided this situation will be remedied and a more healthful and pleasant environment for human habitation will be created.

Excessive soil moisture is often the cause of soil compaction by animals and machines. Adequate drainage permits greater control of the soil water and hence greater ease in the conduct of the farming operations.

A high water table results in a soil that does not warm up readily in the spring. Germination of crops is delayed and the seed may, in fact, rot before it germinates. A drained soil is a warm soil.

A poor root environment exists in areas of high water table. Plant diseases are more active under these conditions. Fungus growth is particularly prevalent.

In irrigated areas the accumulation of salts can convert a fertile area into a barren desert. The economic loss to the community as well as the individual

Figure 1-4 Mosquitos in poorly drained areas.

makes it imperative to consider drainage. Maierhofer, one of the world's leading drainage experts, has written, "Who has seen the salinized lands in the plains of the mighty Indus, so great in extent that 40 million people would not go to bed hungry each night if the potentially high productive lands, now barren and white with salt, were adequately drained?"

HISTORY OF LAND DRAINAGE

The progress of the art of land drainage has proceeded in a cyclic manner. During periods of farm prosperity farmers are eager to get maximum productivity. In addition they have the financial means with which to construct drainage systems. During periods of low farm prices the farmer must restrict his capital outlay and the interest in drainage projects drops very low.

It is known that drainage was practiced in prehistoric times by the Egyptians and the Babylonians. The Greek historian Herodotus in 400 B.C. refers to the use of drainage in the Nile Valley. One of the early recorded references to drainage was by Cato in 2 B.C. He gave the first specific written instructions for land drainage. Pliny, in 1 A.D., noted that covered drains could be constructed by half filling a trench with stones or gravel or a rope "of sprays tied together and the whole covered with earth that had been thrown out."

Columella, in 1 A.D., recommended that the drains be placed 3 feet deep.

The practice of the art of drainage is probably as old as the art of agriculture. The first known examples go back to the Roman Empire and probably earlier. The Romans recognized the importance of soils information as a basis of drainage design, and the superiority of deep and covered drains under certain circumstances. The methods used by these people were little improved until present day tile drainage had its origin in England on the estate of Sir James Graham, Northumberland, in 1810. (An earlier use of tile in France in 1620 in the Convent garden at Maubeuge was not followed by widespread adoption of the practice.)

Now that we are enjoying agricultural prosperity the interest in drainage is probably at an all time high. Farmers and governments interested in obtaining maximum yields are installing many miles of drains each year.

Misconceptions based on a lack of understanding are frequently associated with drainage in the public mind. Both floods and droughts have been blamed on drainage. Still another feature associated with large-scale drainage work is the opposition which stems from a resistance to changing the existing conditions because of a conflict of interest. It is recorded that the drainage of the Fens of Eastern England, comprising over 200,000 acres of land subject to the storm tides of the North Sea, was attended with difficulties and discouragements, the chief of which was the opposition of the fenmen who occupied the lands and derived a precarious livelihood from hunting, fishing, and livestock raising. Present day drainage projects are sometimes subjected to opposition from conservation groups interested in preserving the natural habitats of wildlife.

In spite of the misunderstandings and outright opposition which have occasionally arisen, progress in drainage has resulted in the addition of millions of acres of highly productive land for the production of food and fiber for the ever-increasing numbers of people inhabiting the earth. The fact that the reclamation and preservation of land by drainage has only begun is proved by the observations of many trained agricultural scientists. Large areas in the eastern and southern United States can profit by drainage. The permanence of irrigated agriculture depends, to a degree, on drainage, and many additional miles of drains are needed in arid regions. The United States is not alone in her drainage potential, according to many of our scientists and engineers who travel abroad on foreign missions.

DRAINAGE NEEDS IN ARID REGIONS

It must seem incongruous to many people that irrigated lands frequently require drainage. After heavy initial development costs for irrigation water and land preparation have been paid, why must there be an excess to cause a drainage problem?

Drainage problems have beset irrigators in arid areas since the earliest recorded times in history. The Valley of the Tigris and Euphrates Rivers in old Mesopotamia has largely returned to desert due to the accumulation of salts in the surface soil layers. At one time nearly 10,000,000 acres of land in ancient Chaldea were, to quote the late Sir William Willcocks (Means et al., 1930), "as fertile as a garden." Most of this region now consists of alkali flats and saline areas, barren of all but meager feed for flocks and herds of wandering tribesmen. Relics of abandoned irrigation systems, alkali areas, and saline accumulation throughout the Near East and the Sahara Desert, indicate that lack of proper drainage eventually resulted in economic ruin to the areas. In some cases it probably contributed to the decay and eventual destruction of the civilization which flourished in ancient times (Thorne, 1951). There is evidence in the United States also that early irrigation works constructed by the Indians and later by the Spaniards in the Gila River Valley in Arizona, and in the Rio Grande Valley in New Mexico and Texas eventually had to be abandoned because of problems resulting from lack of drainage. Traces of irrigation systems near the Isleta Indian Reservation, New Mexico, indicate that irrigated crops were raised before the discovery of America (Bloodgood, 1930). Spanish explorers in the sixteenth century found Pueblo Indians practicing a primitive agriculture with the aid of water diverted from the Rio Grande. Much of this land is presently saline and unsuitable for crops in its present state.

One of the most startling things is the rapidity with which drainage problems can develop over large areas after irrigation water is applied. At first glance it would seem that pervious surface soils underlain with sandy layers at shallow depths, having been dry and idle for the centuries, would be safe from waterlogging. Experience has shown the fallacy of this reasoning. The Newlands Project at Fallon, Nevada, was one of the first projects constructed by the Bureau of Reclamation in 1902. A dam was built and irrigation started in 1906 on about 70,000 acres. Waterlogging of the lands began soon after the start of irrigation, and by the end of 1918 more than 35,000 acres of land had the water table less than 6 feet below the ground surface. The construction of deep open drains started in 1921, and by the end of 1923 there were over 150 miles of open drains to carry away both surface waste water from irrigation as well as subsurface waters. Although additional drainage ditches have been provided since then, the area continues to be plagued by the high water table.

Many other instances of the drainage requirements of irrigated areas can be cited. For example, in 1912 there were 65,000 acres of land in the Pecos

Valley, New Mexico, which had been drained. In 1915, R. A. Hart stated that, "Already, in the United States, more than 10% of the entire area that has been irrigated for any considerable period is either absolutely unproductive or is given over to the less valuable crops or to poor pastures. . . . These injured lands are to be found in all the arid and semiarid states and in practically every valley where irrigation is a factor in the agricultural development."

Figure 1-5 Windmill of the type used for draining the lowlands of Holland.

The Soil Conservation Service estimates that 8,000,000 acres of land in 17 western states need improved drainage.

DRAINAGE NEEDS IN HUMID REGIONS

Drainage problems in humid regions differ from those in arid regions. The difference is due to several factors including rainfall, soils, and salt. Surface drainage problems are widespread in humid regions. These problems are the result of surface topography that does not permit the rain to flow easily over the surface of the soil. In many areas the soil cannot absorb the rainfall fast enough and much of it runs off over the surface. Subsurface drainage problems due to high water-table conditions may be caused by water that cannot penetrate dense layers in the soil. These layers may have low permeability to water because of their high clay content, or they may be layers which were compacted by the weight of the ice cap during the last ice age.

In addition there is much land in the humid areas which is subject to periodic inundation due to its position relative to the ocean or other large body of water.

While drainage in humid regions, especially tropical areas, must be very old, one of the most useful and dramatic developments of the art of draining has taken place in Holland. Pliny, in 1 A.D., wrote of the first efforts of Frieslanders to reclaim small bits of land from the sea. However, drainage did not come into its own until the application of the windmill as a means of power for operating drainage pumps in 1408. One of the experts of this era was Jan Leeghwater (1575–1650) who designed important drainage works in Holland and Germany. The most spectacular result of this era was the reclamation of large areas of land from the sea. Most of the concern was with surface drainage, and the important developments in subsurface drainage did not take place until the eighteenth and nineteenth centuries, with the development of clay tile.

The first drain tiles were laid in the United States in 1835 by John Johnstone on a farm at Geneva, New York. These first tiles were handmade. The use of tile received great stimulus with the invention and introduction into the United States of a tile-making machine in 1848. Drains were installed quickly to increase the productivity of the lands along the Eastern Seaboard.

Some idea of the importance of subsurface drainage to the agricultural economy of the United States can be gained when it is considered that in the major states of the corn belt (Iowa, Indiana, Illinois, and Ohio) about 20% of the land is drained, mostly by tile. The cost of tile drains installed in the state of Iowa alone exceeds the cost of the Panama Canal.

REFERENCES

Bloodgood, Dean W. 1930. The ground water of middle Rio Grande Valley and its relation to drainage. *New Mexico Agr. Exp. Sta. Bull.* 184.

Brown, Charles F. 1913. Farm drainage, a manual of instruction. *Utah Agr. Exp. Sta. Bull.* 123.

Clayton, B. S. and L. A. Jones. 1941. Controlled drainage in the Northern Everglades of Florida. *Agr. Eng.*, **22**:287–288, 291.

Gain, Elmer W. 1964. Nature and scope of surface drainage in Eastern United States and Canada. *Trans. Am. Soc. Agr. Eng.*, **7**:167, 168, 169.

Henderson, D. W., J. H. Lindt, Jr. and R. C. Pearl. 1954. Use of moles for subirrigation. *Calif. Agr.*, **8**:5–6.

Hilgard, E. W. 1886. *Report of the California Experimental Station* 1886.

Kalisvaart, C. 1958. *Subirrigation in the Zuyderzee Polders*. Publication No. 2, International Institute for Land Reclamation and Improvement. Wageningen, The Netherlands.

Johnstone, John. 1801. *An Account of the Mode of Draining land According to the System Practised by Mr. Joseph Elkington*. B. McMillan, London.

Land out of the Sea. Published by the Boards of the Zuyderzee Works, The Hague, Netherlands.

Means, Thomas H. 1930. (Chairman) Committee of the Irrigation Division Report. Drainage of irrigated lands. *Trans. Am. Soc. Civil Engrs.*, **94**:1525–1447.

Stuvel, H. J. 1956. *Het Deltaplan*. Scheltema and Holkema N.V., Amsterdam.

Thorne, D. Wynne. 1951. *The Desert Shall Blossom as the Rose*. Faculty Association, Utah State Agricultural College, Logan, Utah.

Van der Molen, W. H. and W. H. Sieben, 1955. *Van Zee tot Land*. W. E. J. Tjeenk Willink N. V., Zwolle, The Netherlands.

van Veen, J. 1955. *Dredge, Drain, Reclaim, the Art of a Nation*. Marinus Nijhoff, The Hague.

Van't Woudt, B. D. and R. M. Hagan. 1957. *Land Drainage in Relation to Soils and Crops*. American Society of Agronomy Monograph 7 (J. N. Luthin, Editor), Madison, Wisconsin.

Weir, W. W. 1954. Land drainage in California. *Agr. Eng.*, **35**:482–485.

Chapter 2 HYDRAULICS AND THE MEASUREMENT OF WATER

THE FLOW OF WATER IN OPEN CHANNELS

Water flows in open channels because of the surface slope of the water. Gravity is the source of energy required to make the water flow. If there is no surface slope there can be no flow. Since gravity is a force per unit mass of the liquid, we know from Newton's law that if there were no resistance to the flow of water it would continue to accelerate just as any body falling freely in a gravitational field. However, the motion of the water in an open channel is opposed by frictional forces between the water and the sides of the channel. Even very smooth channels create frictional forces that oppose the acceleration of the water. For steady-state conditions the water accelerates to a velocity at which the frictional forces are just equal to the acceleration forces. At this point the velocity is steady and does not vary with time.

Since the forces that oppose the flow of water are due to the friction between the water and the walls of the conduit, it is obvious that there will be a variation in the velocity distribution throughout the channel. The surface tension where the water meets the atmosphere also retards the amount of water flowing in the channel. The maximum velocity will occur at the greatest distance from those surfaces which cause resistance to flow. The maximum velocity occurs somewhat below the surface of the water and at the middle of the deepest part of the channel. The distance below the surface depends on the depth of the water. For shallow channels it is near the surface, for deep channels it is about one-third the depth below the surface.

Hydraulic Radius

The resistance to flow is a function of the wetted perimeter and the cross-sectional area of the channel. The area of cross section divided by the wetted perimeter is called the hydraulic radius. For channels having the same area and the same slope, the one having the smallest wetted perimeter will have the largest hydraulic radius and the highest velocity. Experimental work has shown that the relationship between the velocity and the hydraulic radius is such that the velocity varies as the one-half to the two-thirds power of the hydraulic radius.

A number of different formulas have been proposed for calculating the velocity of flow in an open ditch. All of these formulas involve the slope of the ditch, the hydraulic radius of the ditch, and the roughness coefficient of the sides and bottom of the ditch.

The Chezy Formula

One of the first formulas written to evaluate the velocity of water in an open channel is credited to Chezy in 1775. His formula states that

$$v = C(rs)^{\frac{1}{2}} \tag{1}$$

in which v is the mean velocity, r is the hydraulic radius, and s is the slope of the channel. Chezy believed that the coefficient C was a constant, but it has been shown to vary with the roughness of the channel as well as with the velocity and the hydraulic radius. Chezy's formula has been modified to bring it into line with modern concepts of flow in open channels, and the formulas given hereafter are the ones in common use by engineers today.

Manning's Formula

In the United States, Manning's formula is most commonly used for the design of ditches carrying water.

$$v = \left(\frac{1.486}{n}\right) r^{\frac{2}{3}} s^{\frac{1}{2}} \tag{2}$$

where v is the average velocity of flow in the ditch
r is the hydraulic radius
s is the slope of the ditch
n is the roughness coefficient of the ditch.

Since the hydraulic radius occurs to the two-thirds power in Manning's formula, the greatest velocity will occur with the smallest hydraulic radius. For a ditch the smallest hydraulic radius is achieved with a semicircular cross section. It is virtually impossible to maintain such a cross section. The closest practical cross section to the semicircle is the trapezoidal shape. It is this shape that is approximated in the design of a drainage ditch.

The determination of the roughness coefficient is the most difficult task in the use of Manning's formula. It is not a constant factor but will vary from season to season as the sides of the ditch change their shape and as vegetation establishes its growth on the bottom and on the sides of the ditch. In the absence of local measurements the work of C. E. Ramser can serve as a guide in establishing the roughness coefficient to use. Ramser made many measurements of the flow in ditches which had a variety of vegetative growth conditions. Many pictures of these conditions are included in his bulletin, "Flow of Water in Drainage Channels," *U.S.D.A. Technical Bulletin* 129, November 1929.

The experience of the Soil Conservation Service in designing drainage ditches indicates that the following roughness coefficients can be used to give a satisfactory design.

Hydraulic Radius	n
less than 2.5	0.040–0.045
2.5 to 4.0	0.035–0.040
4.0 to 5.0	0.030–0.035
over 5.0	0.025–0.030

The above values are based on the assumption that vegetation in the channels will be kept down by maintenance. The value of n may be 0.100 or higher if vegetation is not kept under control.

In newly dug channels the values of n are lower and the velocities higher than the design values. If it is necessary to reduce the velocities because of erosive conditions, the ditch can be made wider and shallower.

Elliott's Formula

The roughness coefficient can vary a great deal and is sometimes so indeterminate that it is difficult to apply Manning's formula. Because of this difficulty Elliott developed an approximate formula with a constant roughness coefficient. It has the advantage of being easy to use and easy to remember.

According to Elliott the velocity of flow in an open ditch is given by

$$v = (1.5rh)^{\frac{1}{2}} \tag{3}$$

where r is the hydraulic radius and h is the fall or grade in feet per mile.

Kutter's Formula

In 1869, two Swiss engineers, Ganguillet and Kutter, made an analysis of stream flow measurement records and developed an empirical relationship known as Kutter's formula,

$$C = \frac{41.65 + \dfrac{0.00281}{s} + \dfrac{1.811}{n}}{1 + \dfrac{h}{r}\left(41.65 + \dfrac{0.00281}{s}\right)} \tag{4}$$

with C the coefficient in the Chezy formula. The formula expresses C as a function of the hydraulic radius r and the slope s, and the coefficient of roughness n.

The Manning formula is much easier to use than the Kutter formula, and it gives practically the same results except for flat slopes. For flat slopes the Manning formula is more accurate.

THE FLOW OF WATER IN DRAIN PIPES

The flow of water in pipes partially full is due to the slope of the water surface just as it is in open channels. As long as the pipe is not full there will be no pressure head, and gravity will be the sole source of energy causing the water to flow.

Drain pipes rarely run full, and they usually are placed on gentle slopes. Manning's formula for open-channel flow can be used equally well for flow in drain pipes. The resistance factor in a segmented drain pipe is greater than in a smooth continuous pipe. Values for resistance factors and tables for computing the flow in drain pipes are given in Chapter 11, Subsurface Drains.

METHODS OF MEASURING WATER

The measurement of water is important in drainage engineering since it is necessary to know the quantities of water that flow out of a particular drainage basin, or alternatively, it may be desirable to make measurements of the flow from drain pipes. Methods used in measuring water can be grouped into three categories: direct methods, velocity-area methods, and methods which use a cross section of the ditch formed into a constriction to make the measurement.

Direct Methods

The most direct of all methods is the use of a container to collect a given volume in a measured amount of time. For example, an ordinary bucket

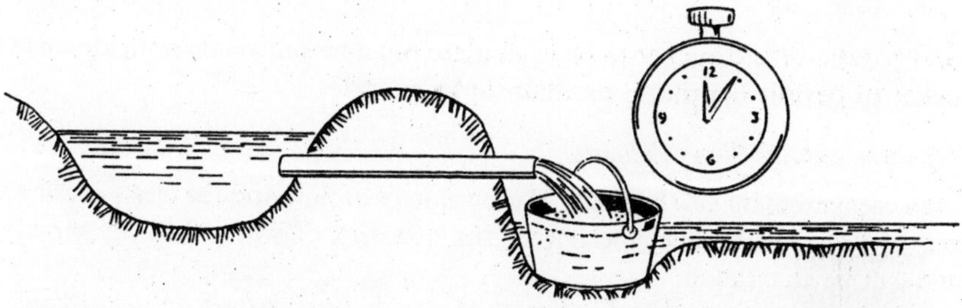

Figure 2-1 Collecting a known volume of water in a measured interval of time.

can be used in conjunction with a stop watch or the second hand of a wrist watch to measure the quantity of water, provided it is sufficiently small. The time required to fill the bucket to the top is measured and from this the quantity rate can be calculated. Another method is the direct measurement of the change of level in the reservoir, or some such gathering place for the water. Deep seepage must be estimated if the measurement is to have any accuracy.

In addition to the use of a container with a stop watch to measure the volume of water, there are a number of commercially available meters which measure the quantity of water. These meters are made primarily to measure the velocity of the water. The velocity is then converted so that the reading on the dial represents the volume of water that is passed through the meter.

An additional method that can be used is the tilt bucket, which consists of a two-compartmented section tank. When one compartment fills to a given volume the tank tilts causing the filled compartment to empty while the other side is brought into the position to fill. A counter mounted on the side of the mechanism adds up the number of tilts, and from this the total volume of water that has passed through the tilting bucket can be calculated. In

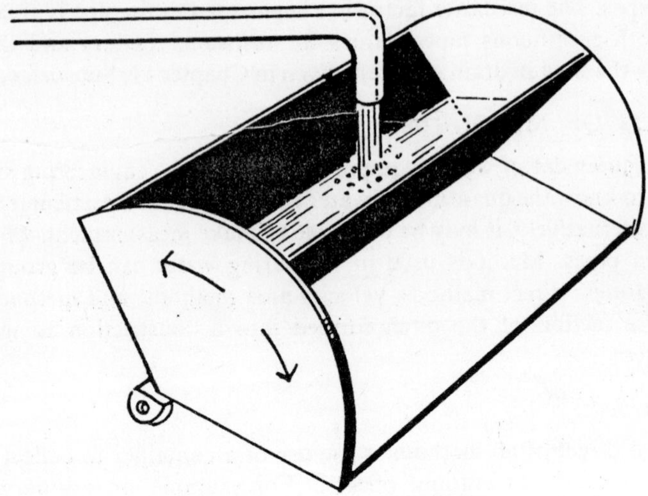

Figure 2-2 Two-way tilt bucket.

order for the tilting bucket to be used there must be sufficient room above the bucket to permit the tilting mechanism to operate.

Measurement of Flow Velocity

If a measurement can be made of the velocity of flow and the cross-sectional area of flow can be also measured, the quantity of water flowing per unit time can be calculated.

THE FLOAT METHOD. The float method can be used to get an approximate measure of the rate of flow. It is useful for estimating the flow in streams. It is necessary to select a straight section of ditch or channel that has a fairly uniform cross section. The length of the section will depend on the current. Usually 50 to 100 feet is adequate. Several measurements of the depth and width should be made within the trial section to obtain an average cross-sectional area. Then a small float, such as a piece of wood, a bottle, or some

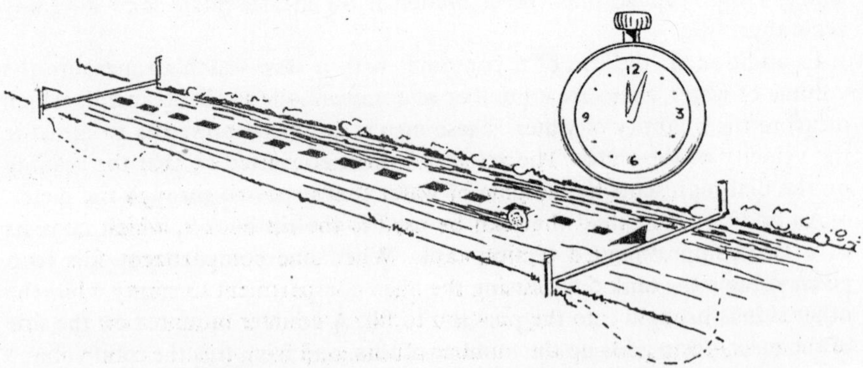

Figure 2-3 Float method for determining rate of flow.

other object, is placed in the ditch at a marked point and the time required for it to traverse the marked section is observed. The measurement should be repeated several times to get an average time of travel. As we noted before, the velocity of flow at the surface will be greater than the average velocity of the stream. It is necessary to multiply the surface velocity by a coefficient of about 0.80 to obtain the average velocity.

THE CURRENT METER METHOD. A current meter is a commercially available device designed to make accurate measurements of the velocity of water. Several types are available. The Price Current Meter contains an impeller

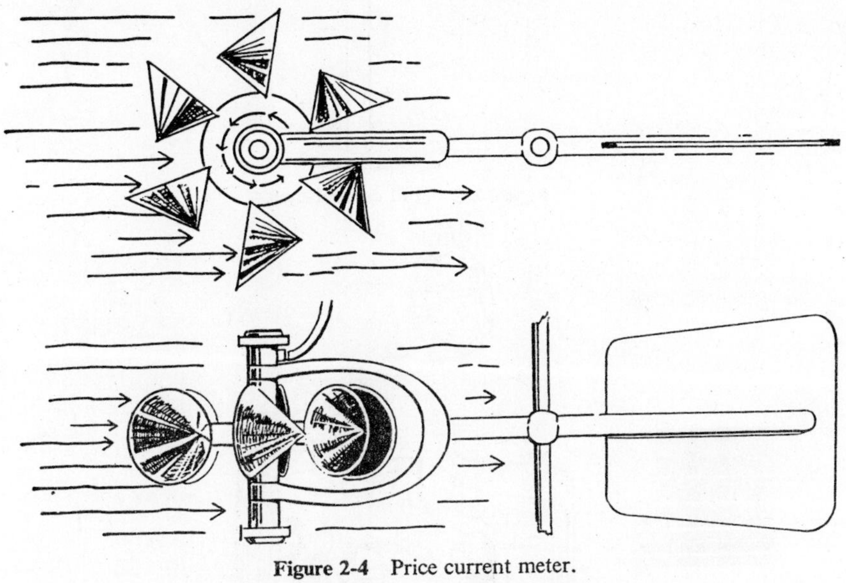

Figure 2-4 Price current meter.

which consists of 6 conical cups mounted on a vertical axis. When the meter is immersed in moving water the impeller revolves and the time for a given number of revolutions is determined by the operator.

The Hoff meter contains a rubber impeller mounted on a horizontal axis. Its chief advantage is that it is less effected by eddies or turbulence. It has been used for measuring the velocity of water flowing from the end of the discharge pipe of pumping plants. For the measurement of velocities in large streams, the current meters are suspended from a cable or walkway which traverses the stream. All of these meters must be calibrated before they are used in the field. A graph or calibration chart is then prepared showing the relationship between the velocity for a given number of revolutions and a given time interval.

To obtain the most accurate information from current measurements, it is necessary that the channel be straight, with a fairly regular cross section. It is desirable to avoid making measurements near piers or other obstructions

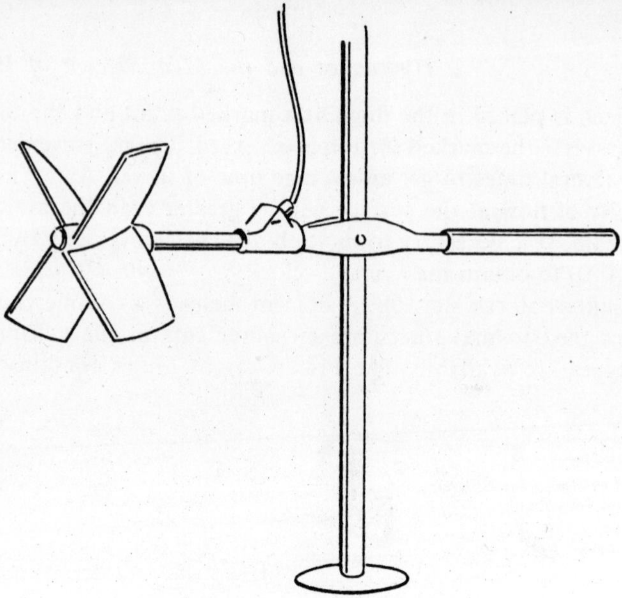

Figure 2-5 Hoff current meter.

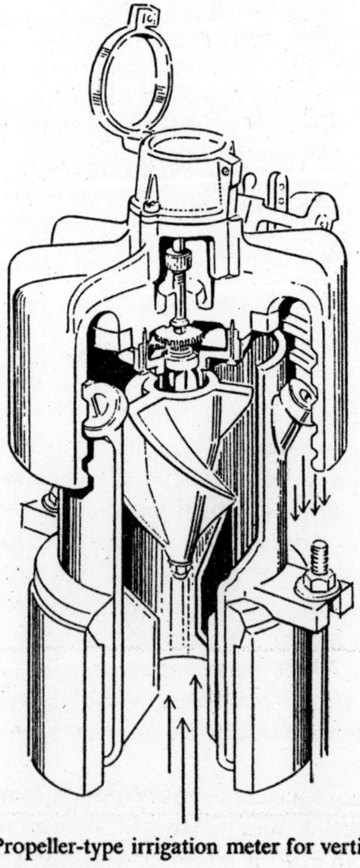

Figure 2-6 Propeller-type irrigation meter for vertical riser pipe.

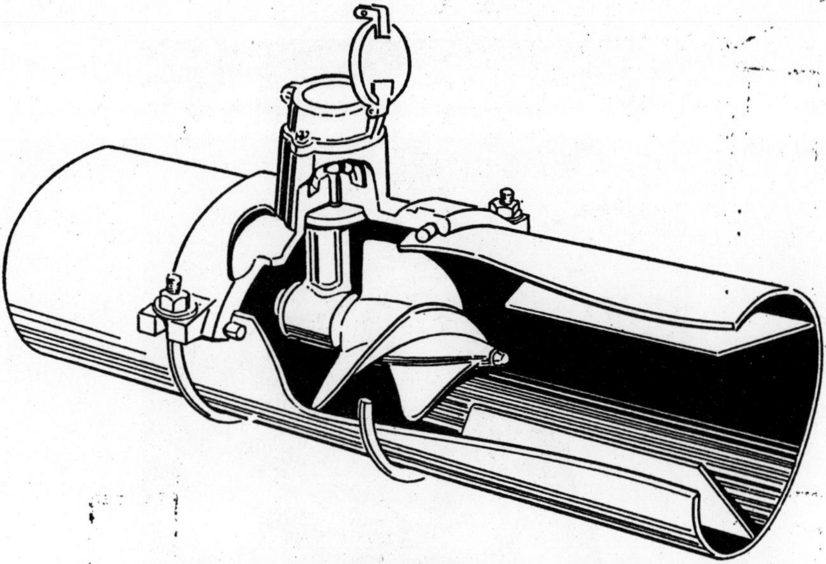

Figure 2-7 Propeller-type irrigation meter within pipeline.

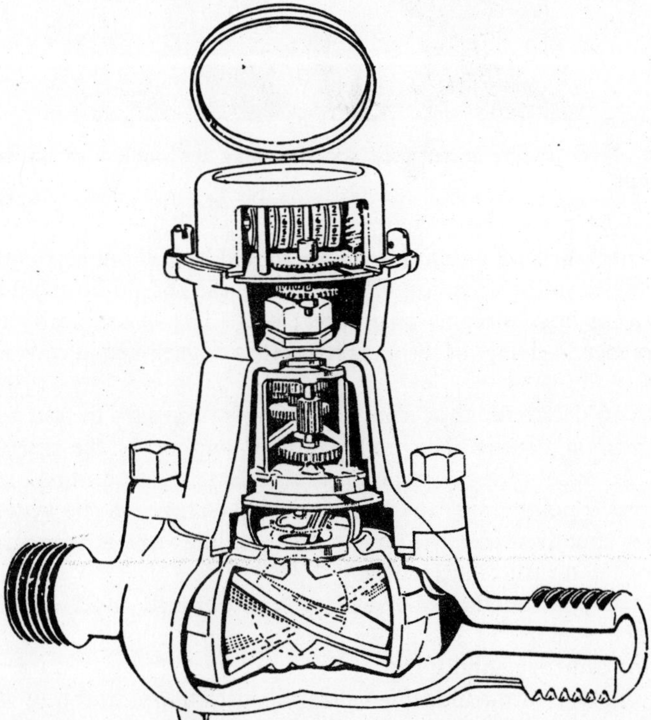

Figure 2-8 Disk-type residential meter.

Figure 2-9 Propeller-type meter used for measuring the outflow of an experimental drainage well.

in the channel. Several measuring points should be laid out across the stream at right angles to the direction of flow, and these should be spaced at equal distances. On large streams an interval of 10 feet is ordinarily used. The depth and mean velocity of the stream are then determined at each measuring point.

In order to determine the mean velocity, it is necessary to take readings at several different depths. The multiple-point method is the most accurate method. At each measuring point the velocity is determined at several closely spaced points from the bottom of the channel to the water surface. If these are equally spaced the mean velocity in the vertical approximates the average of the measured velocities.

In the two-point method the velocity is determined at 0.2 and 0.8 of the depths. The average of these two measurements approximates the mean velocity for ordinary conditions.

In the single-point method the velocity is determined at 0.6 of the stream depth or on the water surface. This method is generally employed at depths less than one foot, a depth which is also insufficient for the two-point method.

In the vertical-integration method the meter is lowered and raised at a uniform rate in each of the selected verticals in the measuring section. Owing to the possibility of introducing errors, this method is seldom used in routine stream gauging.

THE TRAJECTORY METHOD OF MEASURING THE DISCHARGE OF PIPES. The trajectory method can be used to determine the rate of flow from the end of a pipe. If the pipe is flowing full the horizontal and vertical measurements are made on the upper surface of the jet. For partially filled pipes the measurements should be made at the centers of the jets.

THE SLOTTED-TUBE METHOD. The discharge from a drainage pipe can be measured by a recent development called the slotted-tube method. A slotted tube with the slot directed upstream is placed at the end of a discharge pipe.

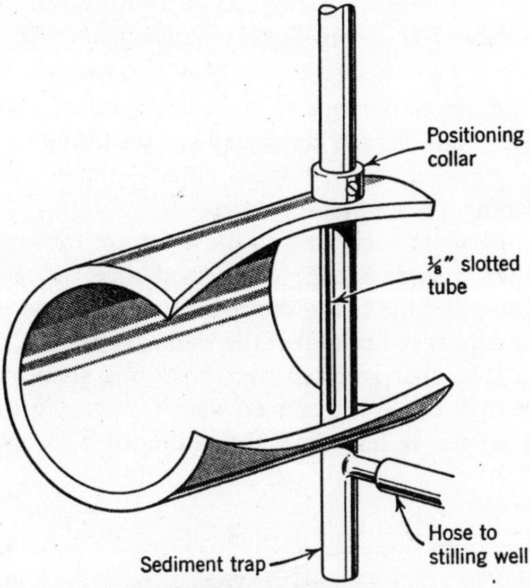

Figure 2-10 Slotted tube.

The arrangement of the parts and the employment of dimensions for the use of the device in smooth-walled pipes is shown in Figure 2-10. The water elevation in the slotted tube is measured from the bottom of the pipe. This number is used as an index of the velocity. For convenience and greater accuracy the slotted tube is connected to a stilling well by a hose. This hose connects to the lower end of the slotted tube. The amount of water in the stilling well can then be measured manually with a tape or a gauge, or by means of a water-stage recorder if continuous records are needed.

Methods Employing a Constriction in the Flow Channel for Making a Measurement

WEIRS. A weir is a notch of regular form through which water may flow. There are a number of different types of weirs, and they are classified according

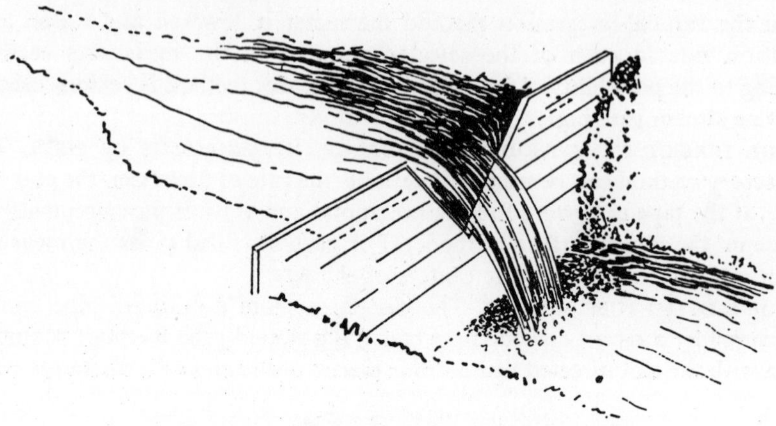

Figure 2-11 Ninety-degree triangular notch weir.

to the shape of the notch, the type of crest, and whether they are contracted or suppressed. For example, the classifications according to the shape of the notch are rectangular contracted weirs, the V-notch weir, the Cipoletti weir, and the rectangular suppressed weir.

Classifications of weirs according to the crest are the sharp-crested weir, which has a sharp upstream edge so formed that the water, in passing, touches only a line; and the broad-crested weir, which has either a rounded upstream edge or a crest so broad that the water, in passing, comes in contact with the surface. The sharp-crested weir possesses greater accuracy and is used much more than the broad-crested weir. Contracted weirs are those in which the crest length is less than the width of the upstream channel.

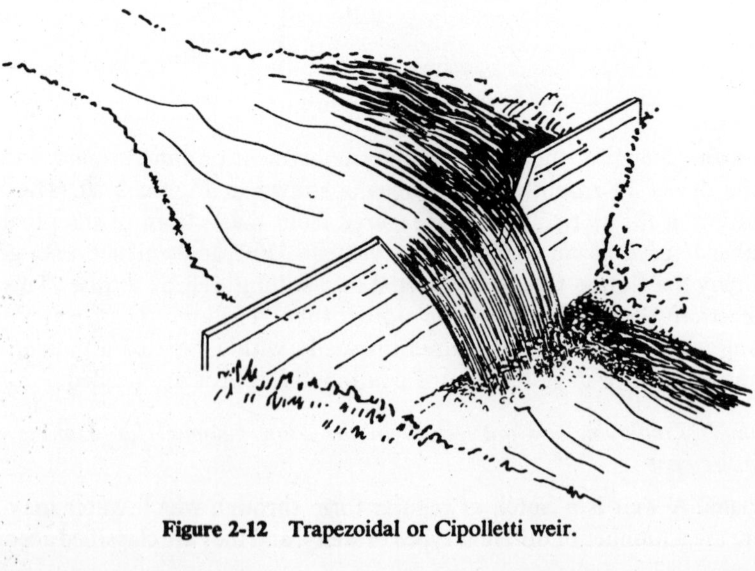

Figure 2-12 Trapezoidal or Cipolletti weir.

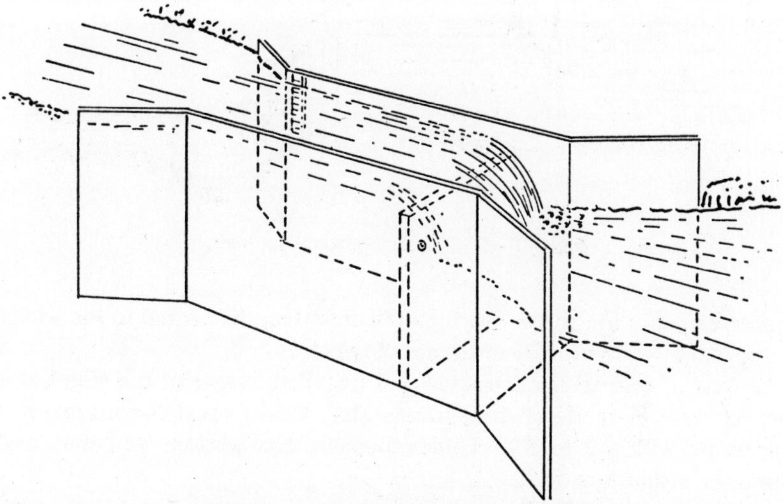

Figure 2-13 Rectangular suppressed weir.

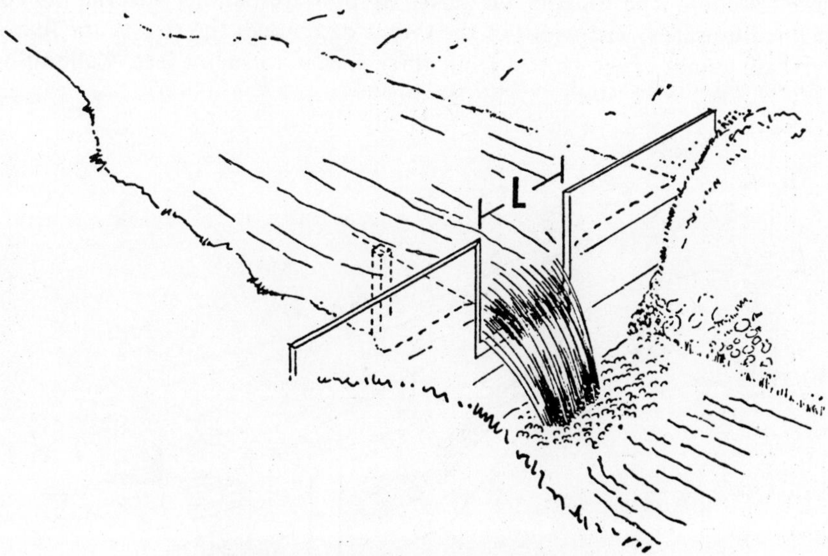

Figure 2-14 Rectangular contracted weir.

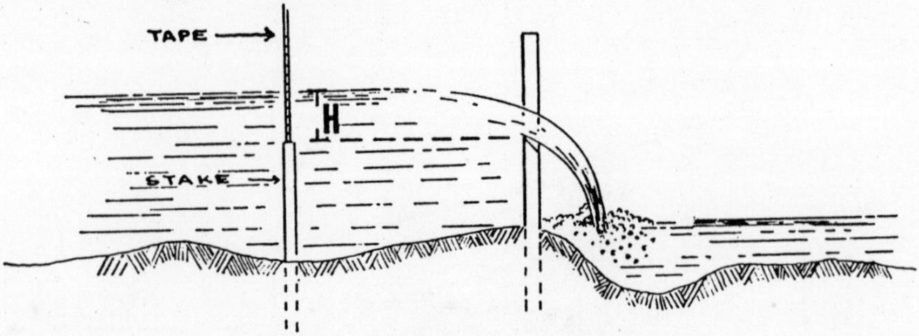

Figure 2-15 Measuring head over weir.

Suppressed weirs are those in which the crest length is equal to the width of the approach section in the upstream channel.

The weir, if properly constructed and installed, is one of the simplest and most accurate methods of measuring water. Under ideal conditions, it will be accurate within 2 or 3%. Under most field conditions you may expect accuracies within 5 to 15%.

PARSHALL FLUMES. The Parshall flume is a self-cleaning flume that operates with a small drop in head. It is widely used in the western United States in irrigated areas.

The floor of the upstream section is level and the walls converge towards a throat section. The walls of the throat section are parallel and the floor is inclined upwards. The width of the throat determines the size of the flume. Parshall flumes range in size from three inches to many feet. Calibration

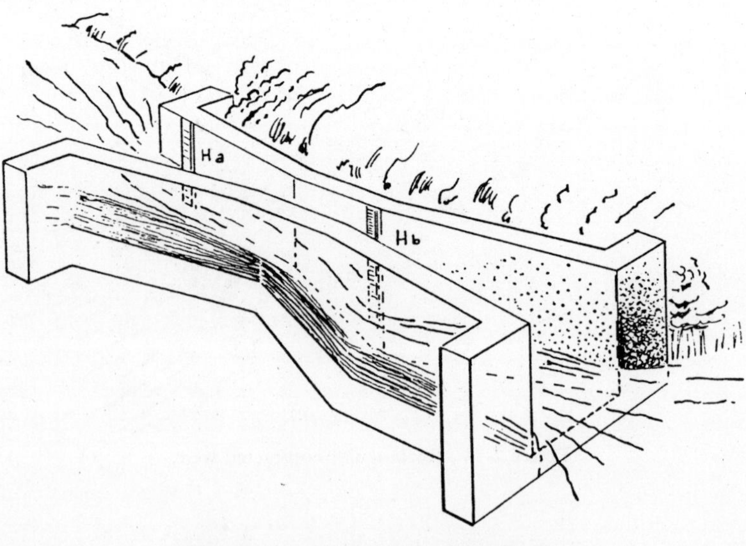

Figure 2-16 Parshall measuring flume.

Figure 2-17 Parshall flume used to measure water in the field.

List of Equivalents

The following equivalents may be used for converting from one unit to another and for computing volumes from flow units:

Volume Units

One acre-inch
- = 3,630 cubic feet
- = 27,154 gallons
- = 1/12 acre-foot

One acre-foot
- = 43,560 cubic feet
- = 325,851 gallons
- = 12 acre-inches

One cubic foot
- = 1,728 cubic inches
- = 7.481 (approximately 7.5) gallons
- weighs approximately 62.4 pounds (62.5 for ordinary calculations)

One gallon
- = 231 cubic inches
- = 0.13368 cubic foot
- weighs approximately 8.33 pounds

Rate of Flow Units

One cubic foot per second
- = 448.83 (approximately 450) gallons per minute
- = 50 Southern California miner's inches
- = 40 California statutory miner's inches
- = 1 acre-inch in 1 hour and 30 seconds (approximately 1 hour), or 0.992 (approximately 1) acre-inch per hour
- = 1 acre-foot in 12 hours and 6 minutes (approximately 12 hours), or 1.984 (approximately 2) acre-feet per day (24 hours)

One gallon per minute
- = 0.00223 (approximately 1/450) cubic foot per second
- = 0.1114 (approximately 1/9) Southern California miner's inch
- = 0.0891 (approximately 1/11) California statutory miner's inch
- = 1 acre-inch in 452.6 (approximately 450) hours, or 0.00221 acre-inch per hour
- = 1 acre-foot in 226.3 days, or 0.00442 acre-foot per day
- = 1 inch depth of water over 96.3 square feet in 1 hour

Million gallons per day
- = 1.547 cubic feet per second
- = 694.4 gallons per minute
- = 77.36 Southern California miner's inches
- = 61.89 California statutory miner's inches

curves are available and tables have been developed for a wide variety of sizes. Stilling wells can be used to measure the water elevations in the flume if turbulence is present or if more accurate measurements are required.

REFERENCES

King, H. W. 1954. *Handbook of Hydraulics*, 4th edition. McGraw-Hill Book Co., New York.

Parshall, R. L. 1950. Measuring water in irrigation channels with Parshall flumes and small weirs. *U.S. Dept. Agr. Circ.* 843.

Rouse, H. 1950. *Engineering Hydraulics*. John Wiley and Sons, New York.

Scott, V. H. and C. E. Houston. 1959. Measuring irrigation water. *Univ. Calif., Div. of Agr. Sci., Circ.* 473.

PROBLEMS

1. Water is flowing to a depth of 2 feet in a trapezoidal ditch having a bottom width of 3 feet and side slopes of 2:1. The roughness coefficient is 0.042. The gradient of the ditch is 0.15. Calculate the velocity using the formulas of Mannings, Kutter, and Elliott.
2. For a pipe flowing partially full the wetted perimeter P is

$$P = \frac{360 - \theta}{360} 2\pi r$$

and the cross-sectional area of flow A is

$$A = \frac{360 - \theta}{360} \pi r^2 + \tfrac{1}{2} r^2 \sin \theta$$

Show that the maximum velocity occurs for $\theta = 57°40'$ and $d = 0.938D$. Use Manning's formual.

3. A trapezoidal open channel of maximum efficiency is to be designed to carry 120 cfs at a velocity of 3 feet per second. What size and shape would you recommend?
4. Calculate the shape and size of a semicircular cross section designed to carry 120 cfs at a velocity 3 feet per second. Compare the wetted perimeter with that in problem 3.

Chapter 3 RAINFALL AND RUNOFF

RAINFALL AND ITS RELATIONSHIP TO DRAINAGE

In humid regions the major source of the water that must be drained from the surface and from the subsurface of the soil originates in precipitation. Some of this precipitation occurs as snow, but the major portion of it is in the form of rain. The amount of rain that falls will play an important part in determining the need for drainage. Not only is the total amount important, but also the rate at which it falls is of significance. The size of the drainage channels which are needed is directly related to the rainfall and to the ability of the ground to absorb the moisture that falls on it. Drainage works which are designed for a certain storm, called the design storm, are usually of adequate size to handle any snow melt that might occur. Therefore the snow can be ignored in the design of the drainage system in most areas.

CHARACTERISTICS OF RAINFALL

Although moisture is always present in the atmosphere, rainfall occurs only under a certain set of conditions. The moisture in the air must be cooled so that it condenses and forms droplets. The moisture condenses on a nucleus, such as a dust particle. The usual mechanism for the cooling of the air to cause precipitation is the lifting of the air mass.

Rainfall is classified according to the methods that cause the rain to occur. There are three different methods whereby an air mass can be lifted to cause cooling and condensation of the atmospheric water vapor. These major types of rainfall are classified as cyclonic, convective, and orographic. As the water vapor cools it condenses to form clouds. If these clouds are further cooled precipitation will be released. If the temperature in the cloud is below freezing snow will fall.

Cyclonic Precipitation

Cyclonic precipitation is caused by lifting of an air mass due to pressure differences. If a low pressure occurs in an area air will flow horizontally from surrounding areas causing the air in the low pressure area to lift. The precipitation that results is called nonfrontal precipitation. If one air mass lifts over another air mass the precipitation is called frontal precipitation. A front is defined as the boundary between two air masses of different temperatures and densities. Cyclonic precipitation can be in the form of drizzle, intermittent rain, or steady rain. If the precipitation is caused by a warm front that rides on top of a cold air mass there will be extensive areas of cloudiness and precipitation. As the warm front approaches a given area the precipitation becomes more continuous and intense. Warm fronts move at a speed of 10 to 30 miles per hour. The rain resulting from a warm front will be in the form of drizzle, intermittent rain, or steady rain.

On the other hand if a cold front occurs as a result of a cold air mass pushing against and beneath a warm air mass, enlarged clouds will develop which will give intense short duration precipitation. These will be in the form of showers and thunderstorms or scattered showers. The cold fronts move at a speed of 20 to 25 miles per hour.

A third possibility is that of an occluded front. An occluded front occurs when a cold front overtakes a warm front. The precipitation pattern is a combination of both warm and cold frontal distributions. Occluded fronts move at a speed of 5 to 30 miles per hour and the rain falls in the form of drizzle, intermittent rain, steady rain, showers and thunder storms, or scattered showers.

Convective Precipitation

Convective precipitation is due to the upward movement of air that is warmer than its surroundings. A thunderstorm that results from the heating of the atmosphere in the afternoon hours is the best example of convective rainfall. It is composed of towering clouds with an anvil composed of ice crystals usually found on top. The vertical air currents develop tremendous velocities and are hazardous to aircraft. Precipitation in the form of showers is of high intensity and short duration.

Orographic Precipitation

Orographic precipitation is caused by air masses rising over mountain barriers or other changes in the land topography. The greatest amount of precipitation falls on the windward slope. The leeward slope often has very little precipitation. Orographic barriers tend to increase both cyclonic and orographic precipitation because of the increased lifting involved. The rainfall is composed of showers and steady rainfall.

The amount of rain that can be expected in any particular storm determines the quantity of water that must be drained from the area. A short intense storm may tax the surface-drainage facilities. However, a longer less intense storm may cause excessive rise of the water table and thus create a serious subsurface drainage problem. It is thus necessary to have data available on the size of storms, their intensity, and the total amount of rain that falls. In many localities rainfall data have been accumulated for many years.

MEASUREMENT OF RAINFALL

In some areas it may be desirable, or even necessary, to get additional information concerning the rainfall. Any open receptacle with vertical sides can be used for the purpose of measuring the precipitation. In the United States the standard gauge developed by the United States Weather Bureau consists of an 8-inch collector for the rain. The rain passes from the mouth of the collector into an inside measuring tube of smaller diameter. The

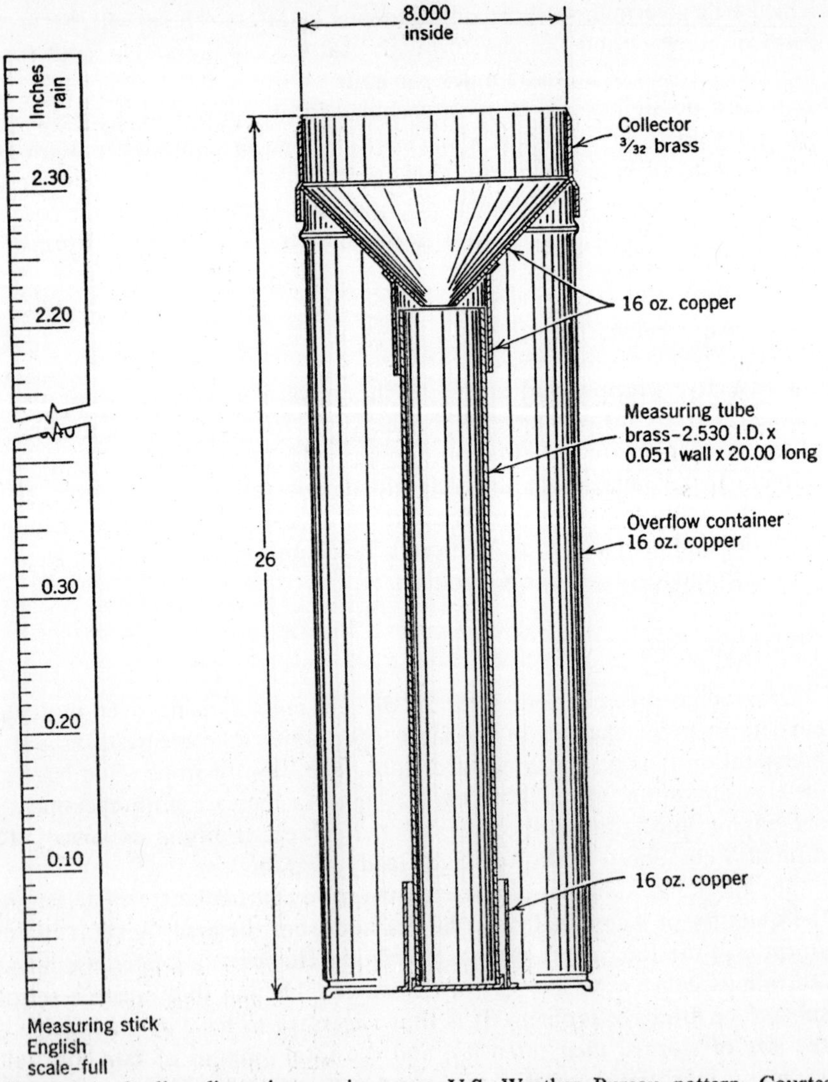

Figure 3-1 Outline dimensions; rain gauge, U.S. Weather Bureau pattern. Courtesy Leupold-Stevens.

measuring tube has a cross-sectional area that is one-tenth that of the collector so that 0.1 inch of rain will measure a depth of 1 inch in the measuring tube.

The most serious error in the collection of rain in a rain gauge results from upward currents of air which cause the precipitation to rise over the gauge.

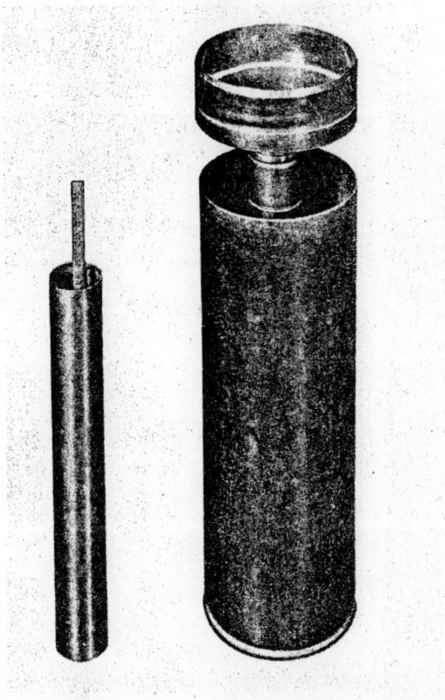

Figure 3-2 Standard Weather Bureau rain gauge. Courtesy Leupold-Stevens.

This error can be reduced by locating the gauge in a sheltered spot adjacent to trees or buildings, provided these are not close enough to interfere with the catch.

RECORDING THE RAINFALL DATA

Rainfall records are valuable only if kept continuously for a period of several years. A single year's record is of very little value in the design of a drainage scheme. For the best results the records should extend over a period of about fifty years. The best records are obtained at stations which are equipped with a continuous rainfall recorder. Not only is the total amount given but also the intensity of each storm can be recorded. A continuous recorder will indicate whether the storm is of short duration and high intensity or of long duration and of low intensity. The rate at which the rain falls is of paramount importance in the design of the drainage system.

Storms of high intensity and short duration are characteristic of the Mississippi Basin in the United States, especially during the summer months.

Figure 3-3 Recording rain gauge. Courtesy Leupold-Stevens.

Many of these intense storms last only one hour. The rate at which the rain falls on the land exceeds the ability of the soil to absorb the moisture, and much of it runs off. The drainage channels must be designed to handle these high intensity storms in order to reduce the crop damage and to prevent damage to structures and improvements on the land. The control of the surface water in these areas is particularly important since the rain falls during the most active part of the growing season when the plants are most susceptible to the effects of flooding.

Low-intensity storms that continue over long periods of time are likely to occur in the coastal areas of the western United States and, to a certain extent, on the Eastern Seaboard. They are also common in most coastal areas of Northern Europe, including England, France, and Holland, as well as neighboring areas.

These storms may last one day or they may last for a week. It is common to have the rain fall for 4 or 5 days. The total amount of rain that falls in any one day may amount to 2 or 3 inches.

Since the rain falls over a longer period of time and at a slower rate, the critical period of runoff is also lengthened. Also there is often less runoff to be handled at any one time. Therefore, the capacity of the ditches and drainage channels need not be so great.

THE MEAN RAINFALL OVER A BASIN OR WATERSHED

The Arithmetical Mean

The arithmetical mean is the simplest method that can be used. It consists of averaging all the amounts that have been recorded at the various stations in the area. If the stations are distributed uniformly over the area and there are not too many variations in the rainfall rate this method is satisfactory. However, if there are differences of topography or other land features which cause variations in the rainfall rate the arithmetical-mean method can lead to large errors.

Thiessen Mean

All of the stations in the area are placed on a map. Lines are drawn between adjacent stations. This divides the area into a series of triangles. On each of these lines, a perpendicular bisector is erected to form a series of polygons, each containing one and only one rainfall station. It is assumed that the entire area within any polygon is nearer to the rainfall station that is included in the polygon than to any other rainfall station. The rainfall recorded at that station is therefore assigned to that polygon. If the mean rainfall on the basin is labeled P, and the area of the basin is A, and $P_1, P_2 \cdots P_n$ represent rainfall records at the stations whose surroundings polygons have the areas $A_1, A_2 \cdots A_n$ then

$$P = \frac{A_1P_1 + A_2P_2 + \cdots + A_nP_n}{A} \tag{1}$$

The Isohyetal Method

A third method is the isohyetal method. Isohyets are contours of equal rainfall. These are drawn on a map after the rainfall at each station is plotted. The area between adjacent isohyets is either estimated or obtained by planimetering the area. The mean rainfall on the basin can be obtained from the equation given above in which A_1, A_2, and A_n are the areas between successive isohyets, and P_1, P_2, and P_n represent the mean rainfall on the respective areas. In drawing the isohyets, consideration must be given to the effect of topography on rainfall rates. An improved map of isohyets will be obtained if some judgment is used to consider the various factors which influence the rate of rainfall. Normally the rain-gauging stations are located in easily accessible areas and very few are found in the mountainous areas.

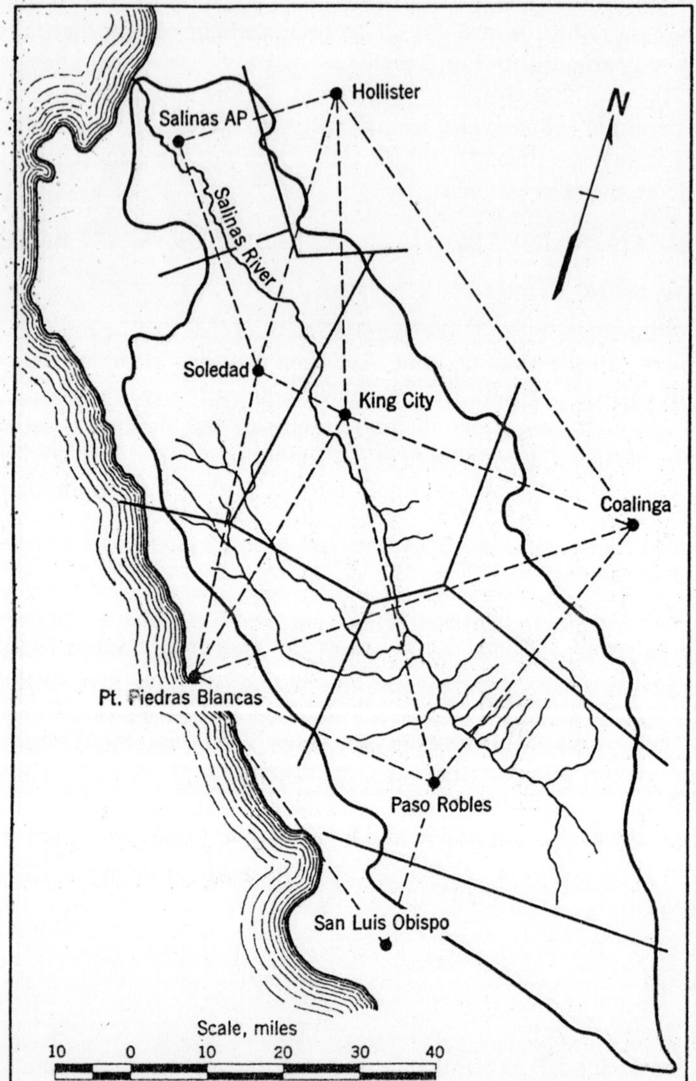

Figure 3-4 Salinas River drainage basin. Thiessen method.

Consideration should be given to the orographic effect of elevation on rainfall in drawing the isohyetal map.

FREQUENCY OF RAINFALL

Structures which are designed to carry storm runoff must be designed for an expected rainfall. Normally, the design is based on the greatest rainfall that will occur with a given frequency. The frequency of occurrence of a

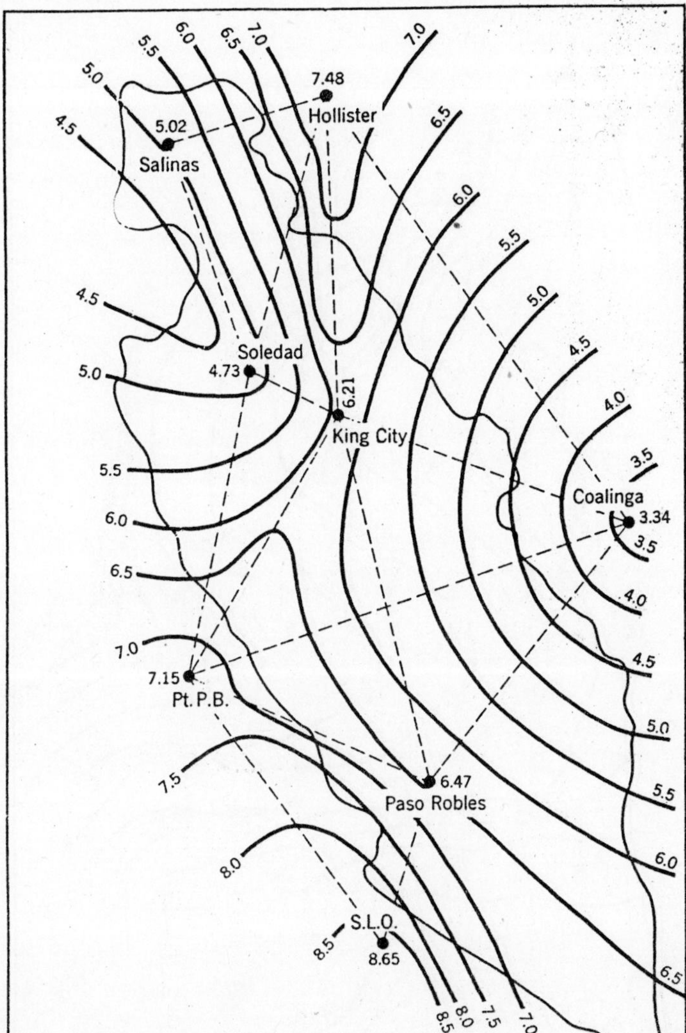

Figure 3-5 Isohyetal method.

rainfall represents the number of years in which a rain of that magnitude, or larger, will occur. It is the average recurrence interval of rains equal to or greater than a certain magnitude. For example, a 10-year frequency for one-hour rain is that magnitude of hourly rainfall which can be expected to be equaled or exceeded ten times in 100 years. It does not mean that such rains will occur at 10-year intervals. It is quite likely that rains of that magnitude will occur oftener than every 10 years, and in fact, two such rains may occur

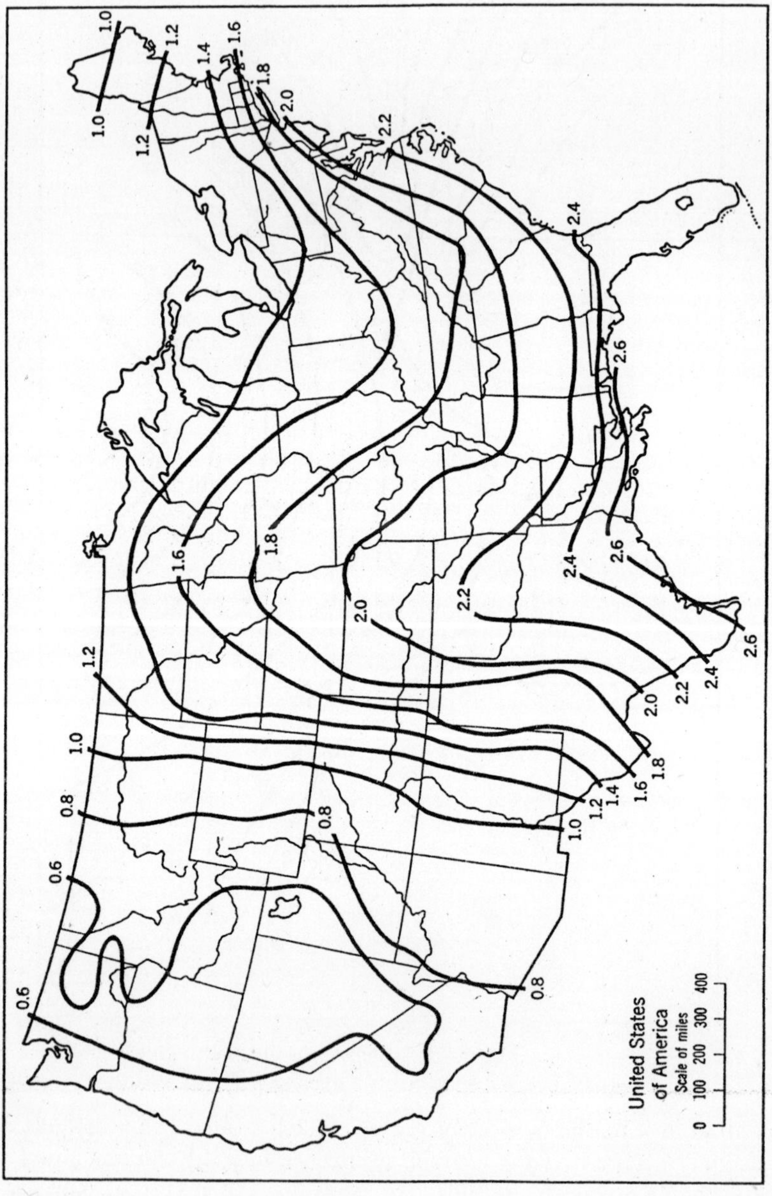

Figure 3-6 Thirty-minute rainfall expected to occur once in 10 years.

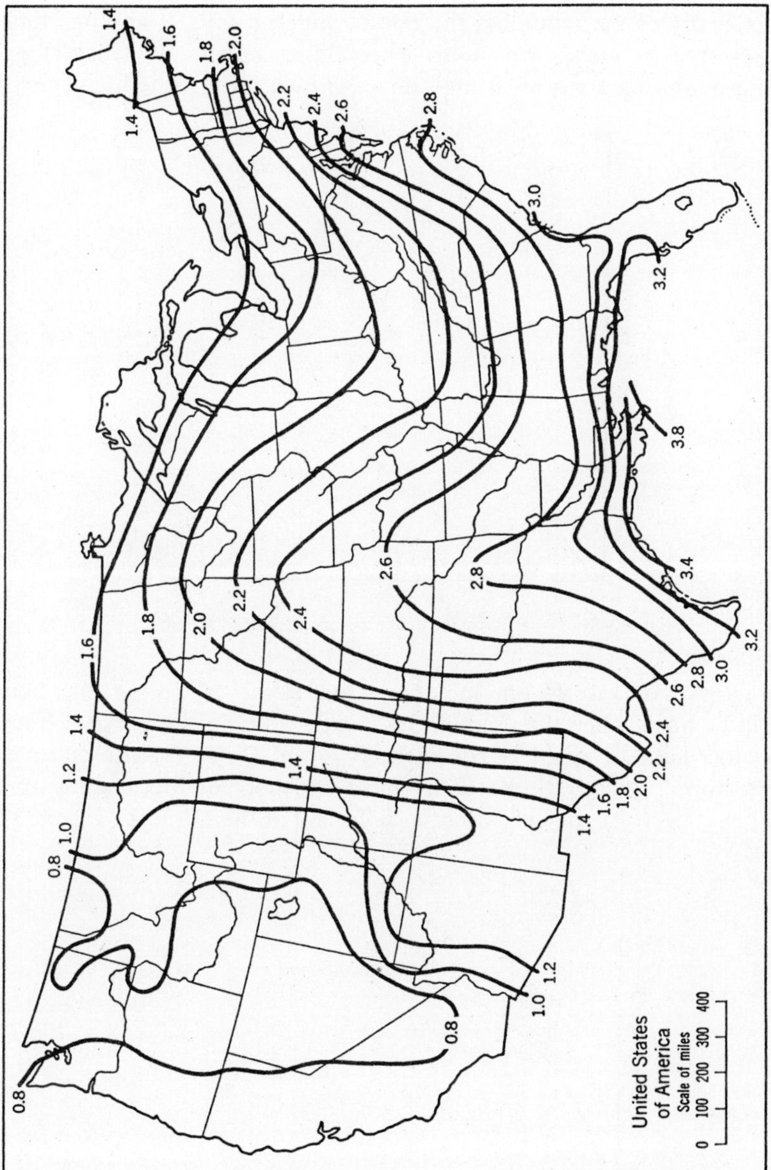

Figure 3-7 One-hour rainfall to be expected once in 10 years.

in the same year or even in the same month. The recurrence interval in this illustration is 10 years.

CHARACTERISTICS OF A STORM

The intensity of the rainfall is the rate at which it falls at any one time. It is expressed in inches per hour. The rate at which rain falls changes continuously during a storm. It may rain one inch during one hour, giving

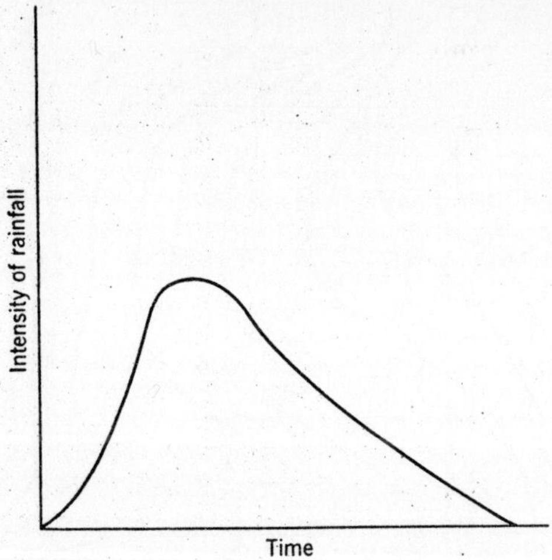

Figure 3-8 Time-intensity curve for an idealized storm.

an average rainfall rate of one inch per hour. However, during that hour there will be times when the rainfall rate greatly exceeds one inch per hour. On the other hand, it might be considerably lower. Figure 3-8 illustrates an idealized storm. Initially the intensity increases more or less linearly to a peak. Then it decreases at a slower rate to the end of the storm. In Figure 3-9

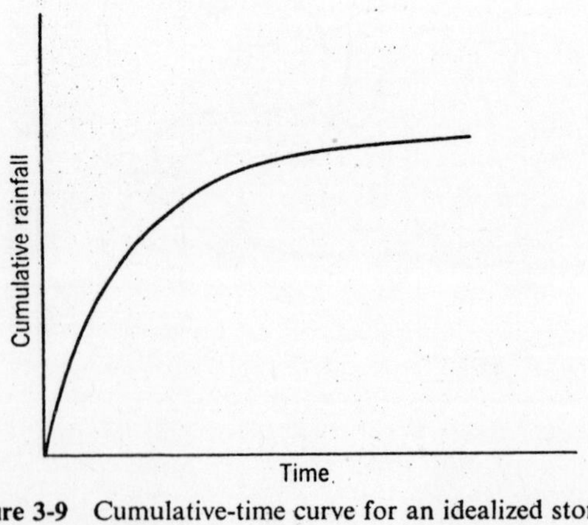

Figure 3-9 Cumulative-time curve for an idealized storm.

the total cumulative amount of water is plotted against time. The curve rises steeply at the start of the storm and falls off towards the end of the storm.

THE DESIGN STORM

In surface drainage the structures and facilities are designed for a storm of a certain intensity and duration. The capacities of the drains must be adequate to handle the anticipated runoff for a given storm. The problem is to decide on the storm to be used in the design. The first step is to tabulate all available data on rainfall. From these data the most intense storms are selected. Data for each severe storm are then tabulated to indicate the most severe rain that fell during a particular time interval. In other words the 5-minute duration is the most rain that fell during a 5-minute interval during the storm. The 10-minute duration is the maximum amount of rain that fell during any 10-minute period of the storm.

The next step is to tabulate the data for the severe storms according to their severity. Table 3-1 illustrates a tabulation for a number of storms. Note that the 5–, 10–, or 15–minute rainfall is not taken from only one storm but represents the maximum for all storms. If we take each of these maxima, the result is called a composite storm, since it consists of data from several storms.

It is now necessary to consider one of these composite storms for the purpose of design. The storm that is selected is called the design storm. The design storm selected is based on the premise that one such storm might be expected to occur once in a given number of years. If the design storm is based on the following formula it is assumed that there is an even chance (one to one) that such a storm, or one more severe, will occur in a set period of years

$$Y = ab \tag{2}$$

where Y is the number of years that rainfall records have been kept, a is the design storm frequency, and is the number of years in which the design storm might be expected to occur; b is the ranking of the severe storm.

As an example, suppose that the rainfall records have been kept for a period of 42 years. It is desired to pick the storm that will be expected to appear once in 10 years, so that $a = 10$. Since $b = Y/a$, then $b = 42/10 = 4.2$. Since the ranking of the storm is 4.2 we would select the fourth most severe composite storm for the purpose of design. There will be more severe storms, but they would be expected to occur more rarely.

The data from the composite storm which is to be used for design purposes are then plotted. The intensity duration is plotted as a function of time. The intensity of rain at one hour is the arbitrary value that is used to compare to standard rainfall-intensity curves. Values of rainfall intensity for periods other than one hour are then obtained from the standard curves. For example, suppose the one-hour intensity is 1.8 inches per hour. The 1.8 curve is used

TABLE 3-1 Tabulated Rainfall Data: Excessive Storms for 31 Years, 1904 to 1934 Inclusive

Duration	5 Minutes		10 Minutes		15 Minutes		30 Minutes		60 Minutes		90 Minutes		120 Minutes	
	Year	Precipitation in inches	Year	Precipitation in inches	Year	Precipitation in inches	Year	Precipitation in inches	Year	Precipitation in inches	Year	Precipitation in inches	Year	Precipitation in inches
U.S.	1908	0.85	1908	1.20	1908	1.40	1908	1.74	1908	2.15	1908	2.46	1915	2.97
Weather	1921	0.76	1915	1.04	1915	1.18	1904	1.55	1904	1.92	1915	2.38	1908	2.63
Bureau	1915	0.73	1921	0.93	1904	1.11	1915	1.36	1915	1.70	1904	2.14	1904	2.34
Records	1934	0.72	1904	0.88	1921	1.03	1921	1.22	1926	1.45	1921	1.81	1921	2.12
	1929	0.66	1926	0.84	1926	0.97	1926	1.18	1921	1.40	1926	1.65	1926	1.83
	1926	0.62	1934	0.80	1934	0.92	1931	1.10	1914	1.33	1914	1.50	1917	1.64
	1931	0.51	1929	0.78	1929	0.90	1934	1.05	1931	1.25	1931	1.40	1914	1.55
	1904	0.45	1931	0.68	1931	0.82	1929	1.01	1934	1.20	1917	1.36	1931	1.51
	1917	0.36	1911	0.52	1911	0.67	1911	0.95	1929	1.14	1934	1.34	1934	1.46
	1914	0.28	1917	0.51	1917	0.62	1917	0.83	1911	1.11	1929	1.27	1929	1.41
	1911	0.21	1914	0.39	1914	0.50	1914	0.79	1917	1.09	1911	1.23	1911	1.34

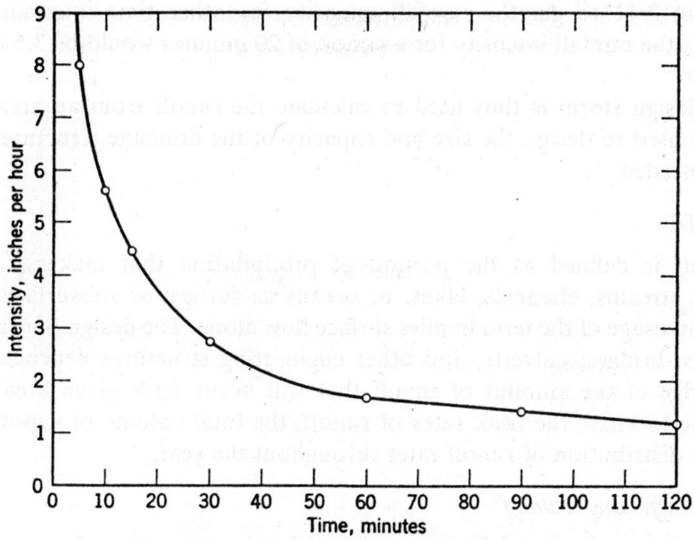

Figure 3-10 Intensity-duration curve.

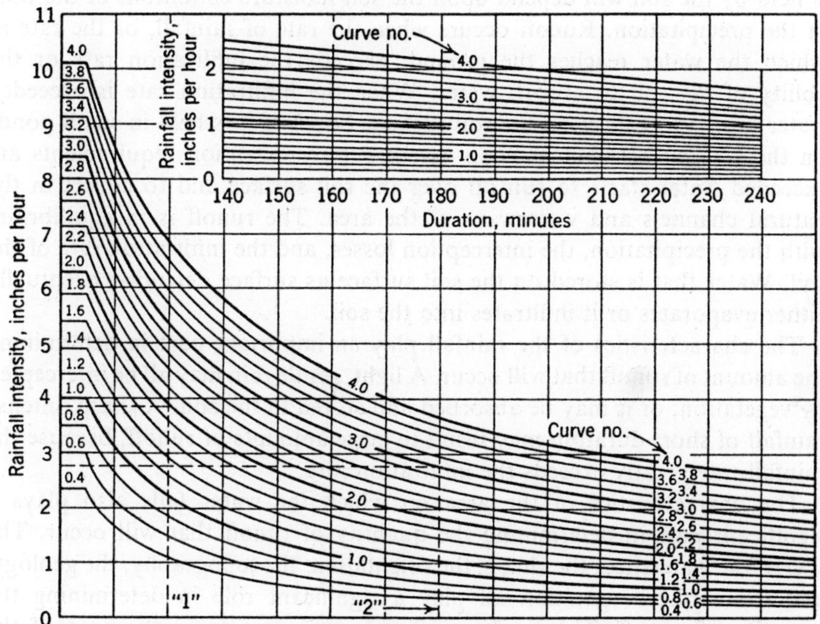

Figure 3-11 Standard intensity-duration curves. Note: Curve numbers correspond to one-hour values of rainfall or supply indicated by respective curves. All points on the same curve are assumed to have the same average frequency of occurrence.

in Figure 3-11 to get the rainfall intensity for other time intervals. For example, the rainfall intensity for a period of 20 minutes would be 3.5 inches per hour.

The design storm is thus used to calculate the runoff from an area, and hence is used to design the size and capacity of the drainage structures that will be needed.

RUNOFF

Runoff is defined as the portion of precipitation that makes its way towards streams, channels, lakes, or oceans as surface or subsurface flow. Common usage of the term implies surface flow alone. The design of drainage channels, bridges, culverts, and other engineering structures depends on a knowledge of the amount of runoff that will occur on a given area. It is desirable to know the peak rates of runoff, the total volume of runoff, and also the distribution of runoff rates throughout the year.

Factors Affecting Runoff

Part of the rain that falls is intercepted by the vegetation. Some of it is stored in depressions on the ground surface and is called surface detention. Some of the precipitation is absorbed and held by the soil. The amount that is held by the soil will depend upon the soil-moisture conditions at the time of the precipitation. Runoff occurs when the rate of rainfall, or the rate at which the water reaches the ground, exceeds the infiltration rate or the ability of the soil to absorb water. When the infiltration rate is exceeded water begins to collect on the soil surface. Some of it is held in small ponds on the soil surface, but as soon as the surface-detention requirements are exceeded water starts to run off over the soil surface and to collect in the natural channels and waterways of the area. The runoff is in equilibrium with the precipitation, the interception losses, and the infiltration rate of the soil. Water that is stored on the soil surface as surface detention eventually either evaporates or it infiltrates into the soil.

The characteristics of the rainfall play an important part in determining the amount of runoff that will occur. A light, gentle rain may all be intercepted by vegetation, or it may be absorbed and stored in the soil. A sharp, intense rainfall of short duration may result in large amounts of runoff, because the rainfall rate greatly exceeds the infiltration rate.

The characteristics of the area on which the water falls also plays a significant part in determining the quantity of runoff that will occur. The size of the watershed, the shape, the orientation, the topography, the geology, and the surface vegetation all play a significant role in determining the quantity of runoff. The total runoff volume increases as the area of the watershed increases. However, the runoff rate does not increase in the same proportions. The volume per unit area of the watershed decreases as the area of the watershed increases.

The shape of the watershed also has a role. A long, narrow watershed is likely to have lower runoff rates than a more compact watershed of the same size. The orientation of the watershed to the storm path is also important. If the long axis of the watershed is parallel to the storm path, a storm moving upstream will cause a lower peak-runoff rate than storms moving downstream.

As an example of some interception losses, some data obtained by the United States Department of Agriculture indicate that a continuous stand of alfalfa intercepted about 0.08 inch of water, while corn, soybeans, and oats intercepted 0.04, 0.04, and 0.01 inch respectively.

The effect of soil on infiltration can be seen in the accompanying data where the Ruston sandy loam absorbed only 0.04 inch, and the other soil types ranged from 0.29 to 2.47 inches.

The condition of the soil also plays an important part. If the soil has been plowed up and down the slope, there is a tremendous increase in the percentage of runoff and a corresponding increase in soil erosion. With Marshall silt loam on an 8% slope, for example, the runoff factors increases from 0.1 to 10.3% in terms of percentage of runoff, and the soil loss increased from 0.0 to 11.75 tons per acre. Houston black clay on a 4% slope shows an increase of 5.77 to 13.13 tons per acre.

The Rational Formula for Estimating Peak Runoff Rate

For small watersheds the rational method is widely used for estimating the peak runoff rates. In its simplest form

$$Q = CiA \tag{3}$$

where Q is the design peak runoff rate in cubic feet per second, C is the runoff coefficient, i is the rainfall intensity in inches per hour for the design recurrence interval and for the duration equal to the time of concentration of the watershed, and A is the watershed area in acres.

The Time of Concentration (TOC)

The time of concentration of the watershed is the time required for water to flow from the most remote part of the area to the outlet. It is assumed that when the duration of the storm equals the time of concentration all parts of the watershed are contributing simultaneously to the discharge at the outlet. For example, if a storm starts at 8 A.M., and at 8:20 A.M. the entire area is contributing water to the outlet or to some particular structure for which a design flow is needed, the time of concentration, called the TOC, is equal to 20 minutes. As the storm begins, if we assume that raindrops fall evenly over the whole area, sooner or later the rain will fall faster than it can infiltrate the soil and water will begin to flow down the slopes of the area. In Figure 3-12 is shown a culvert at the low corner of a rectangular drainage area that slopes upward to point A. The rain which falls at point B immediately enters the culvert and goes through. However, the drop which falls at point A at the same time is just starting the journey towards the culvert. As soon as it leaves the corner of the field another raindrop takes its place and a

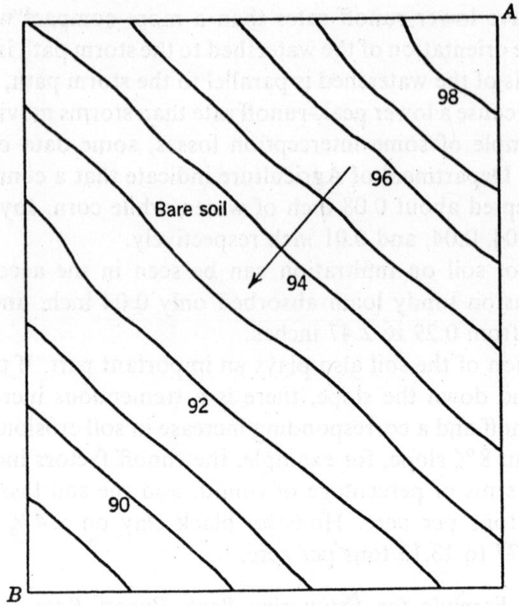

Figure 3-12 The time of concentration is the time required for a drop of water to travel from *A* to *B*.

second drop falls on *A*. Drop *A* keeps moving until at the culvert it joins a recently fallen drop from *B* and, together with the accumulation of drops which have fallen on it during the journey, enters the culvert. At this moment the entire area can be considered to be contributing water to the culvert. Until then the area at point *A* had not contributed to the flow through the culvert.

There are a number of different individual times of flow that influence the time of concentration. For example, the time required for water to enter a ditch or drainage channel, and the time required for water in the ditch or drainage channel to reach the point being considered. Another factor is the time of overland flow which is abbreviated as TOF.

The Time of Overland Flow (TOF)

The time of overland flow varies with the slope, the type of surface of the ground, the length of the flow path, and a number of complicating factors. In Figure 3-13 some values are given for times of flow for various conditions of cover, slope, and length. These are the values used by the Corps of Engineers, and other government agencies, in design work. If the drainage area consists of several different types of surfaces, the time of overland flow must be determined by adding together the respective times computed for flow over lengths of different surfaces along the path and from the most remote point to the inlet.

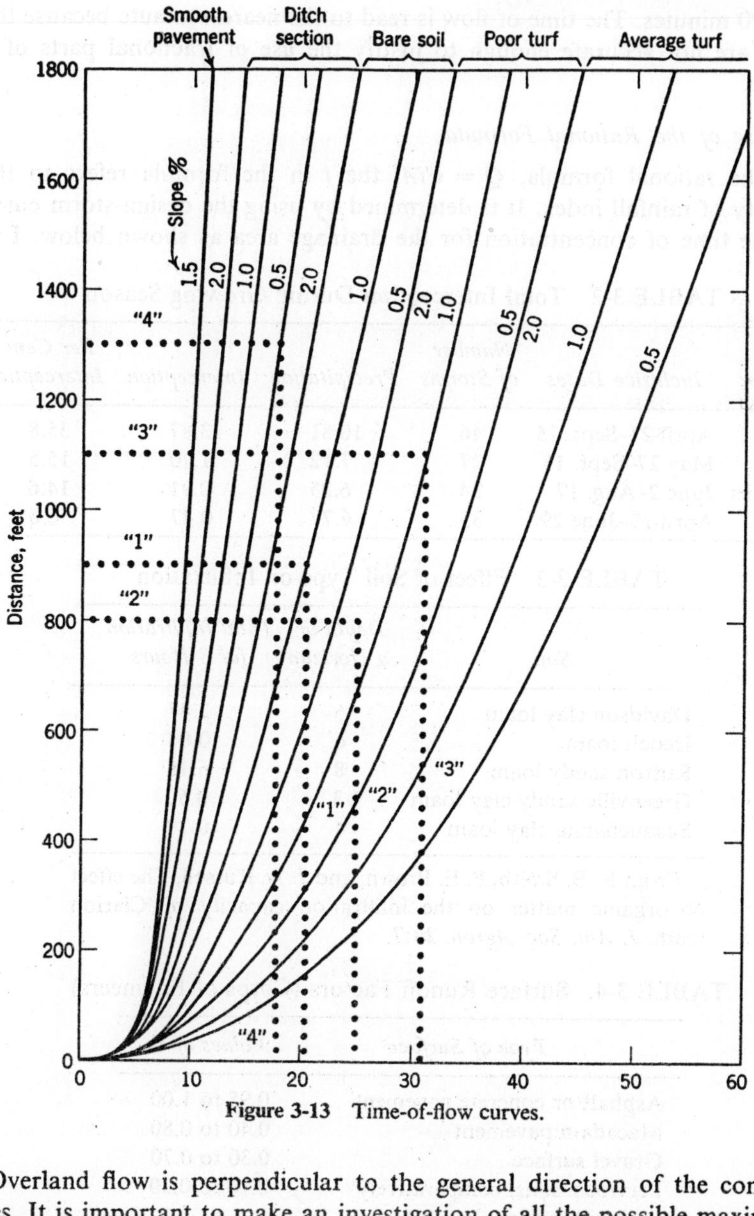

Figure 3-13 Time-of-flow curves.

Overland flow is perpendicular to the general direction of the contour lines. It is important to make an investigation of all the possible maximum lengths of flow so that a realistic value of the time of concentration can be obtained.

EXAMPLE OF DETERMINING THE TIME OF CONCENTRATION. Let us assume that in Figure 3-12 the average slope between points *A* and *B* is 1.0%, the distance of flow is equal to 900 feet, and the area consists of a bare soil. As shown by the dotted line, No. 1, on Figure 3-13, the time of concentration is estimated to be 20 minutes. The time of flow is read to the nearest minute because the charts are not accurate enough to justify the use of fractional parts of a minute.

The Use of the Rational Formula

In the rational formula, $Q = CiA$, the i in the formula refers to the intensity of rainfall index. It is determined by using the design-storm curve and the time of concentration for the drainage area as shown below. For

TABLE 3-2 Total Interception During Growing Season

Crop	*Inclusive Dates*	*Number of Storms*	*Precipitation*	*Interception*	*Per Cent Interception*
Alfalfa	April 27–Sept. 15	46	10.81	3.87	35.8
Corn	May 27–Sept. 15	27	7.12	1.10	15.5
Soybeans	June 2–Aug. 17	24	6.25	0.91	14.6
Oats	April 15–June 29	35	6.77	0.47	6.8

TABLE 3-3 Effect of Soil Type on Infiltration

Soil	*Depth of a Horizon*	*Total Infiltration for 3 Hours*
Davidson clay loam	6	2.47
Iredell loam	6	0.04
Ruston sandy loam	8	6.18
Greenville sandy clay loam	3	0.60
Susquehanna clay loam	4	0.29

From F. B. Smith, P. E. Brown, and J. A. Russell, The effect of organic matter on the infiltration capacity of Clarion loam. *J. Am. Soc. Agron.* **29**:7.

TABLE 3-4 Surface Runoff Factors (Corps of Engineers)

Type of Surface	*Values of C*
Asphalt or concrete pavement	0.95 to 1.00
Macadam pavement	0.40 to 0.80
Gravel surface	0.30 to 0.70
Pervious soils, comparatively flat, heavy turf	0.10 to 0.30
Pervious soils, comparatively sharp surface slope, sparse turf	0.30 to 0.70
Wooded areas (depending on surface slope and soil cover)	0.02 to 0.30
Rocky, barren soils (depending on surface slope, soil cover, etc.)	0.30 to 0.80

example, given a design storm of 2 inches per hour and a time of concentration of 20 minutes, the problem is to determine i for use in the rational formula. For this purpose one uses the standard rainfall-intensity duration curves. Enter the abscissa at 20 minutes and proceed vertically to intersect the curve, then proceed horizontally to the ordinate and read i as 3.9 inches per hour.

REFERENCES

Horn, D. L. and G. O. Schwab. 1963. Evaluation of rational runoff coefficients for small agricultural watersheds. *Trans. Am. Soc. Agr. Eng.*, 6:195–198, 201, 233.

Thiessen, A. H. 1911. Precipitation averages for large areas. *Monthly Weather Review*, July, p. 1082.

U.S. Department of Agriculture. 1941. Climate and man. *Yearbook of Agriculture*, 1941.

U.S. Weather Bureau. 1955. Rainfall intensities for local drainage design in the United States. *Tech. Paper* 24, Part I.

U.S. Weather Bureau. 1955. Rainfall intensity-duration-frequency curves. *Tech. Paper* 25.

Wisler, C. O. and E. F. Brater. 1959. *Hydrology*. John Wiley and Sons, New York. 408 pp.

Yarnell, David L. Rainfall intensity-frequency data. *U.S. Dept. Agr. Misc. Publ.* 204.

PROBLEMS

1. (a) Prepare intensity-duration curves for a once in 2-, 5-, and 10-year composite storms from the data presented in the figure below. The records have been kept for 22 years.

Duration in Minutes						
5	10	15	30	60	90	120
0.84 in.	1.27 in.	1.45 in.	1.78 in.	2.19 in.	2.53 in.	2.82 in.
0.74	1.14	1.27	1.58	1.95	2.30	2.60
0.71	1.00	1.15	1.38	1.78	2.15	2.40
0.70	0.93	1.03	1.22	1.46	1.82	2.10
0.68	0.83	0.94	1.15	1.35	1.60	1.84
0.64	0.80	0.91	1.13	1.32	1.54	1.68
0.50	0.75	0.90	1.08	1.26	1.43	1.56
0.48	0.69	0.83	1.00	1.23	1.34	1.52
0.34	0.56	0.66	0.94	1.18	1.31	1.46
0.29	0.53	0.64	0.84	1.10	1.25	1.40
0.24	0.39	0.46	0.78	1.06	1.21	1.33
0.23	0.36	0.45	0.68	1.05	1.20	1.31
0.21	0.35	0.43	0.65	1.01	1.19	1.25
0.19	0.32	0.39	0.64	0.98	1.09	1.18

(b) Use the standard rainfall-intensity curves to obtain the rainfall intensities at 20, 90, and 150 minutes.

2. Obtain local rainfall records and plot an isohyetal map.
3. Use the Thiessen method to calculate the rainfall at a location.
4. Obtain rainfall-intensity records and prepare a listing of the most severe storms.
5. Compute the average rainfall from the data below by the Thiessen method. Compare results with arithmetical mean.

Rain Gauge	Area in Acres	Rainfall in Inches
1	46.5	1.78
2	13.8	2.15
3	23.6	1.96
4	17.2	2.32
5	12.0	1.85
6	29.2	1.62

Chapter 4 SOILS

DEFINITION OF SOILS

The word soil means different things to different people. The engineer considers soil to be material which supports foundations, roads or airports. The engineering definition of the term soil is very broad. Soil is considered to be all of the material which covers the rock of the earth's crust. The soil scientist or pedologist regards soil as that part of the earth's crust which has been changed as a result of the soil-forming processes. He is concerned with the surface mantle of material subject to the forces of weather and climate. His interest is limited to the first few feet of material to a depth of 5 to 10 feet.

In agricultural drainage work the interest is not only in the soil that forms the surface mantle as defined by the soil scientist, but also the deeper strata that are beyond the reach of the factors that cause soil formation. The primary interest is in the surface layers that support plant growth but we are also interested in the deeper soil layers that can transmit water.

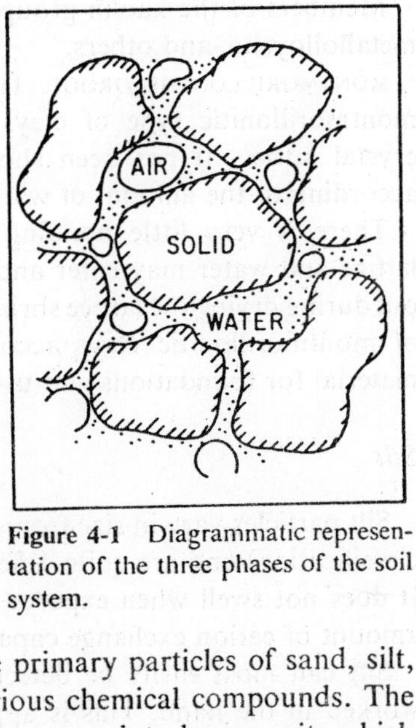

Figure 4-1 Diagrammatic representation of the three phases of the soil system.

THE SOLID PHASE OF SOILS

Soil consists of more than the solid material we can see and feel. It consists of a gaseous phase called the *soil air* and a liquid phase called the *soil moisture*. Each of these phases plays an important part in the use of soils for plant growth and in the engineering use of soils.

The solid phase of the soil consists of the primary particles of sand, silt, and clay along with organic matter and various chemical compounds. The primary particles are usually bound, in various ways, into larger units which are grouped under the general term of soil structure.

Let us discuss first of all the primary particles and something of their nature and behavior.

Clay

Probably the most important single group of particles are the clays. They are less than two microns in diameter and play an important role in soil fertility as well as influencing gross soil properties such as permeability and the mechanical behavior of soils. Clays are complex aluminosilicates. The predominant elements are aluminum and silicon bound together in a crystalline structure. The structure of clays can be determined by X-ray analysis, and it has been possible to distinguish several different types of clays. These clays differ not only in their structure but also in their properties. It is well to consider some of their characteristic properties since they may have a large effect on the way in which a soil drains.

KAOLIN GROUP. Clays of the kaolin group have fixed crystal lattices. They may be regarded as a stacking of thick sheets (7 Å) of layers of oxygens, silicons, oxygens and hydroxyls, and aluminums and hydroxyls. The successive layers are held firmly together by hydrogen bonds.

The kaolin minerals exhibit only slight hydration; hence changes in moisture content cause only small swelling and shrinkage effects. Base exchange or cation exchange is also low, and many of the kaolinitic soils are low in fertility as well.

Members of the kaolin group include kaolinite, dickite, nacrite, endellite, metaholloysite, and others.

MONTMORILLONITIC GROUP. The important property that characterizes the montmorillonitic type of clays is caused by absorption of water in the crystal lattice. It has been shown that the lattice expands and contracts according to the amount of water present.

There is very little bonding force between the adjacent sheets of the lattice and water may enter and cause it to swell. As the water is squeezed out during drying the lattice shrinks. Because of the expansion and contraction of montmorillonitic clays according to the water content they are poor material for foundations and usually have low permeabilities.

Silt

Silt particles vary in size from 0.005 to 0.05 mm (in the unified engineering classification) and are quite different from clay. Silt is generally nonreactive. It does not swell when exposed to water, nor does it have any appreciable amount of cation exchange capacity.

Silt can most easily be detected by a slipperiness of the soil when it is worked in the hand. This is apparently due to the plate-like nature of the individual grains.

The ease with which the silt particles can slide over one another when wet makes it an undesirable material for construction purposes. Because of its instability it is difficult to maintain open ditches in silty soils. Also, silt may move into tile drains and cause them to be plugged.

Sand

Sand particles may be as coarse as rock salt, as fine as powdered sugar, or any size between those extremes. The presence of sand in a soil is noted by a grittiness and roughness when a moist ball of soil is worked between the fingers.

Very fine sands can sometimes exhibit an instability much like that of silt. It is often difficult to maintain open ditches in very sandy soils.

SOIL PORES

The soil pores are the spaces among the solid particles. These solid particles may be individual grains or they may be aggregates of the individual particles. It is in the soil pores that we find the other two phases for our soil system: the soil air and the soil water.

Total Porosity

The total volume of the soil pores is called the porosity. It is usually expressed as a percentage.

If we let n be the porosity then

$$n = \frac{\text{volume of air} + \text{volume of water}}{\text{volume of air} + \text{volume of water} + \text{volume of solids}} \times 100$$

Soil structure plays an important part in determining the porosity, although porosity is also closely related to the size of the primary particles.

Sandy soils may have porosities in the range of 35 to 50% while clayey soils may be from 40 to 60%.

In other words the more finely divided are the individual soil particles the greater is the porosity.

The *void ratio*, e, is defined as the ratio between the volume of pores and the volume of the solids. It is always expressed as a decimal and may have a value greater than one. It may range from as low as 0.30 for a very dense granular material to 2.0 or higher for some clay soils.

The *porosity* and, more usually, the *void ratio* are used in engineering work concerned with compaction of foundation and subgrades. There is no direct relationship between the porosity and the soil permeability.

The *bulk density* is the weight of the soil divided by its volume.

Pore-Size Distribution

Of far greater significance than the porosity is the pore-size distribution, since the size of the pores influences the passage of water and air through the soil. In a later section it will be shown how a knowledge of the pore size distribution can be used to calculate the soil permeability.

More will be said about the use of the capillary tube hypothesis and the determination of the soil pore-size distribution in the chapter on soil water.

SOIL CLASSIFICATION

Engineers Classification

There are a number of different ways in which soil can be classified. The engineer classifies soil on the basis of those characteristics which determine how a soil will behave as an engineering-construction material. One of the several systems of soil classification in use today by engineers is called the unified soil-classification system. The unified soil-classification system recognizes the fact that natural soils rarely exist separately as sand, gravel, or any other single component, but are usually mixtures. Each component contributes its characteristics to the mixture and helps to determine how it will behave as an engineering-construction material. Some of the characteristics of soils which are important in determining the engineering properties are particle size, gradation, particle shape, density, and consistency. Those of interest to us here are explained as follows.

PARTICLE SIZE. Soils are made of grains of particles which range in size from fairly large masses of rock to submicroscopic materials or clay. In the unified soil-classification system, the coarse gravel particles are comparable in size to a lemon, an egg, or a walnut, while fine gravel is about the size of a pea. The size of sand particles ranges from that of rock salt through table salt or granulated sugar to powdered sugar. The particles which pass a No. 200 sieve are called *fines*. Fines are divided into silt and clay.

The physical process of separating the soil into its particle-size groups is known as mechanical analysis. The soil is first passed through a series of

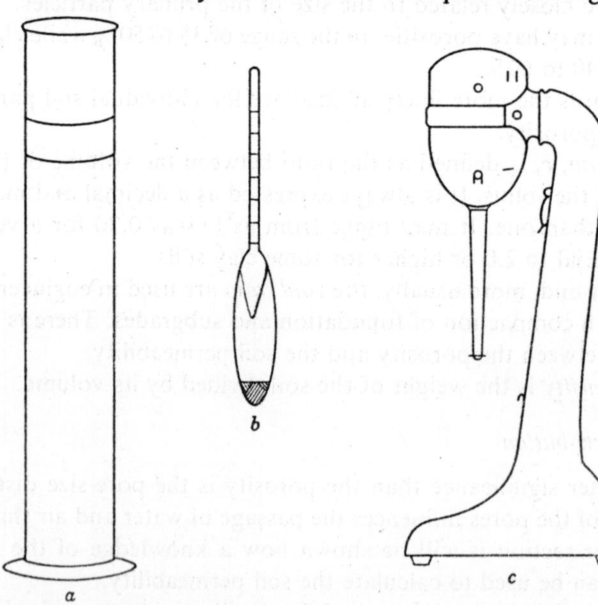

Figure 4-2 Apparatus for the hydrometer method of analysis of the particle-size distribution of a soil. (*a*) Graduated cylinder, (*b*) hydrometer, (*c*) stirrer.

sieves to separate the particles larger than silt size. The sieve analysis consists of separation of the soil into its fractions by shaking the dry, loose material through a nest of sieves of increasing fineness, that is, with successively smaller openings. The sieve analysis may be performed directly upon soils which contain little or no fines. However, if there are some fines in the soil, this will cause aggregation of the coarser particles, and this aggregation must be destroyed by crushing before the analysis is performed. The practical lower limit for the use of sieves is a No. 200 sieve which has openings which are 0.074 mm square and has a total of 40,000 openings per square inch.

It is often desirable to determine the distribution of particle sizes that are smaller than 0.05 mm. These may be determined by a process known as the wet mechanical analysis, which employs the principle of sedimentation. It takes advantage of the fact that the rate at which particles fall through a liquid is dependent upon the diameter of the particle. The two common methods employed are the pipette method and hydrometer method, developed by Bouyoucos in 1927 and Casagrande in 1934. The results of the mechanical analysis are presented in several different ways. The engineer presents the results in the form of a *grain-size* distribution curve. The most commonly used method of presenting the results of the mechanical analysis graphically is shown in Figure 4-3. The particle diameters or sieve sizes are plotted horizontally on a logarithmic scale of a semi-log plot, while the percentage by weight of the material passing any given sieve size, or smaller than any given grain size, is plotted vertically on the arithmetic scale. These points are then connected to form a smooth curve which is called the grain-size distribution curve. The percentage corresponding to a particle size of less than 0.02 mm is determined by wet mechanical analysis.

This method of reporting has two principal advantages. First, it permits ready visualization of the distribution of particle sizes or gradation. Second, it permits the determination of percentages which correspond to particles which were not included in the original mechanical analysis.

GRADATION. The distribution of particle sizes in a soil is known as its gradation. The grain size which corresponds to 10% on a grain-size distribution curve is called *Hazen's effective size* and is designated by the symbol D_{10}. The effective size of clean sands and gravels can be related to their permeability and is, therefore, of importance in drainage engineering.

The *uniformity coefficient* is defined as the ratio between the grain diameter that corresponds to 60% on the curve, and D_{10}.

The coefficient of gradation is given by the expression

$$Cg = \frac{(D_{30})^2}{D_{60} \times D_{10}}$$

Well-graded soils are those which have a reasonably large spread between the largest and the smallest particles, and have no marked deficiency in any one size.

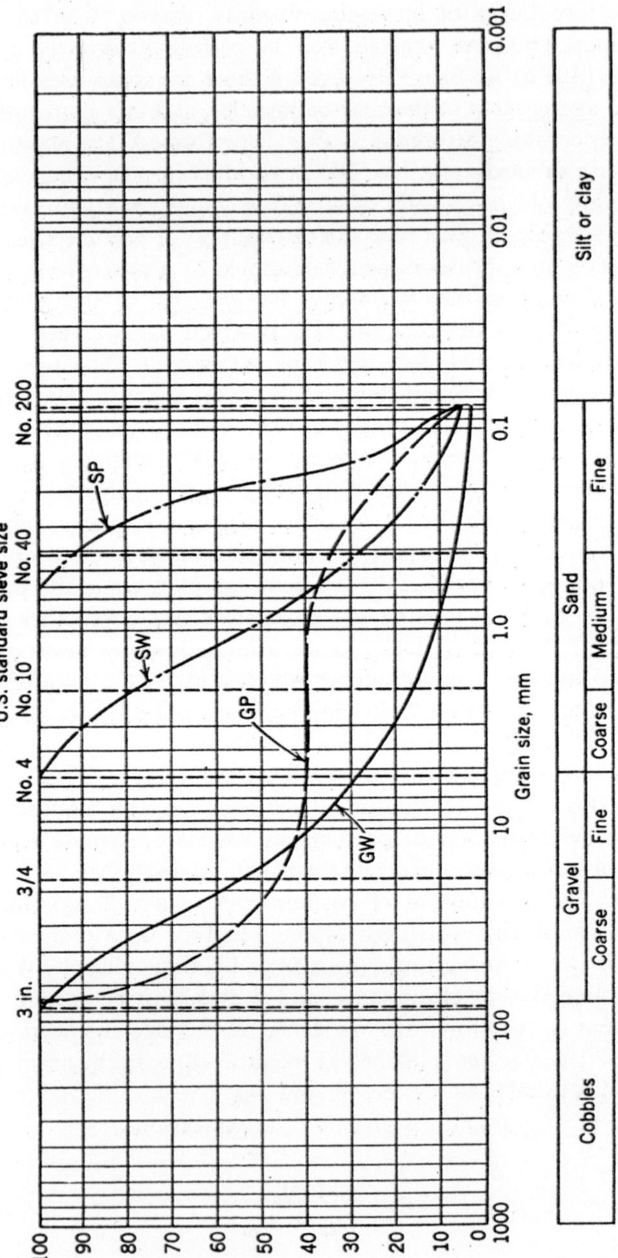

Figure 4-3 Typical grain-size distribution curves for well-graded (*W*) and poorly graded (*P*) soils.

TABLE 4-1 Soil Classification—Corps of Engineers

Sieve or Screen	*Sieve Opening*, mm
2 in.	50.8
$1\frac{1}{2}$ in.	38.1
1 in.	25.4
$\frac{3}{4}$ in.	19.1
$\frac{1}{4}$ in.	6.35
No. 4	4.75
No. 10	2.00
No. 40	0.42
No. 60	0.25
No. 100	0.149
No. 200	0.074

Cobbles	greater than 3 in.
Coarse gravel	$\frac{3}{4}$–3 in.
Fine gravel	4 mesh to $\frac{3}{4}$ in.
Coarse sand	10 mesh to 4 mesh
Medium sand	40 mesh to 10 mesh
Fine sand	200 mesh to 40 mesh

Silt and clay will pass the 200 mesh screen.

Pedologists' Classification

THE SOIL PROFILE. The soil profile is a vertical section through the soil mass. The soil scientist classifies soils according to naturally formed bodies. This classification is done by a careful examination of the soil profile. The factors that influence soil formation—climate, topography, organisms, parent material, and time—usually act on the top 3 to 6 feet of the soil mass.

In drainage we may be interested in the soil at considerable depths, but the surface layers are usually the most important in determining drainage design. Hence, the soils maps prepared by the pedologists can be of utility to us.

According to the pedologist the soil profile can be roughly divided into two horizons, the horizon of *eluviation* and the horizon of *illuviation.*

The zone of eluviation is the zone which is being leached. Chemicals and clays are being dissolved and carried to greater depths in the soil. Soil scientists refer to the zone of eluviation as the A horizon.

Illuviation refers to the zone in which these chemicals or clays are being deposited and are accumulating. Soil scientists call this the B horizon. This zone is sometimes referred to as the subsoil.

Below the A and B horizons lies the parent material or the raw unweathered material from which the soil has been formed. If the parent material is rock then the soil is referred to as residual soil or primary soil. If the material has been transported by wind or water then the soil is referred to as a secondary or transported soil.

The "age" of a soil is a measure of its development. The greater the accumulation of clay or solid chemicals in the B horizon the "older" is the soil. In actuality this accumulation depends on other factors in addition to time. The amount of rainfall and other climatic factors, the topography of the land and its surface drainage pattern, and the influence of plants and soil microbes all play a part in the aging of the soil.

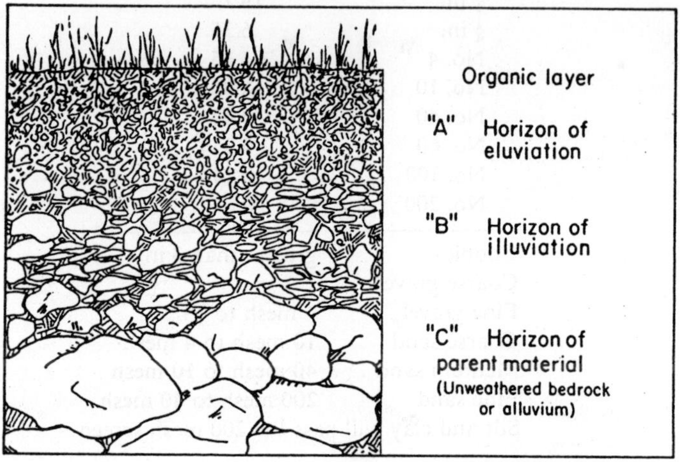

Figure 4-5 A soil that has formed in place (residual soil).

In arid regions many of the soils are very "young." The effects of stratification then become important.

Stratified soils may have clayey subsoils that have no relation to the age of the soil. Stratification can be recognized by the abruptness of the change from one stratum to another. In a developed soil the transition from the A to the B horizon is usually gradual.

Soil Texture

The solid part of the soil can be separated into primary particles. Since the primary particles are bound together by inorganic chemicals as well as organic compounds some treatment is necessary to remove the bonds that hold these particles together.

The relative proportion of sand, silt, and clay determines the soil texture. Texture can be regarded as the "feel" of the soil when it is kneaded in the hand and worked with the fingers, which is the common method of its determination in the field.

Soil texture can play an important part in the design of drainage systems since, in some areas, it is possible to relate texture to soil permeability.

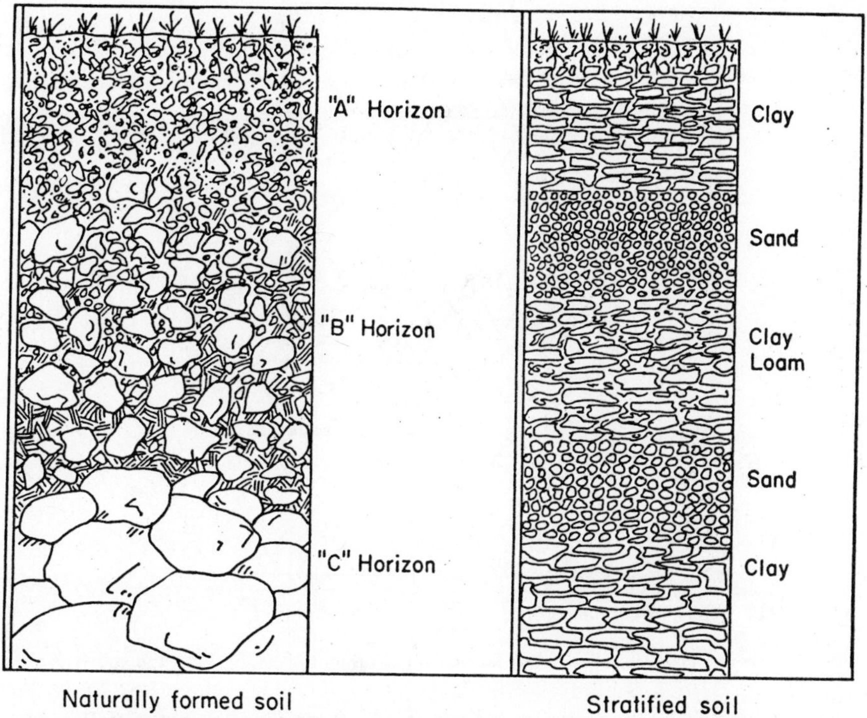

Figure 4-6 A naturally formed soil and a layered or stratified soil.

The Soil Series

Soils which have common properties are grouped together into soil series. A name is assigned to each series and this name is usually a geographic place name of a locality associated with the soil.

The Soil Type

The textural classification of a soil is called the soil type, and a complete identification of the soil is made by specifying both the series name as well as the type. For example a Yolo clay loam is a soil having a clay loam texture and belonging to the Yolo series. A complete description of the series can be obtained from the soil-survey reports of the various state experiment stations or from other governmental agencies charged with the responsibility of making soil surveys.

The size limits proposed by the International Society of Soil Science are:

Separate	*Diameter Limits*, mm.
Coarse sand	2.00–0.20
Fine sand	0.20–0.02
Silt	0.02–0.002
Clay	below 0.002

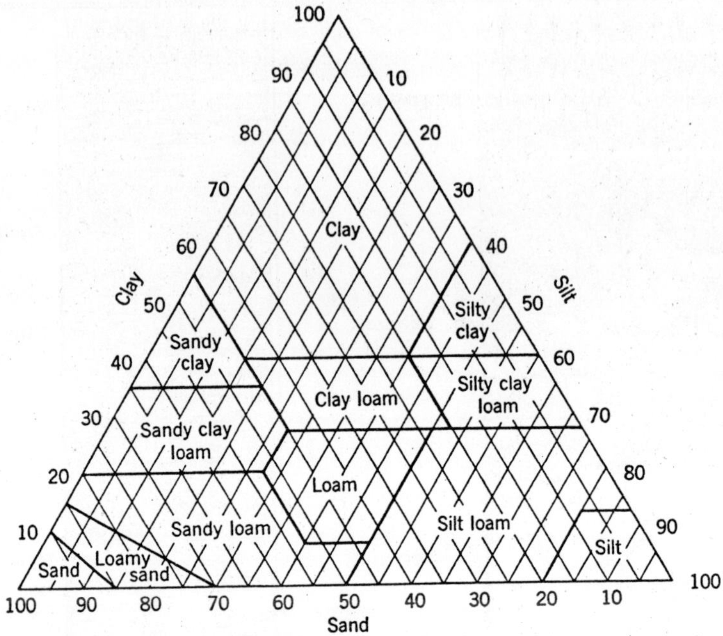

Figure 4-7 Textural triangle used by the soil scientist for establishing the soil type.

The United States Department of Agriculture proposes the following classification:

Separate	*Diameter Limits*, mm
1. Very coarse sand	2.00–1.00
2. Coarse sand	1.00–0.50
3. Medium sand	0.50–0.25
4. Fine sand	0.25–0.10
5. Very fine sand	0.10–0.05
6. Silt	0.05–0.002
7. Clay	below 0.002

REFERENCES

Baver, L. D. 1956. *Soil Physics*, Third Edition. John Wiley and Sons, New York.

Bouyoucos, G. J. 1927. The hydrometer as a new method for the mechanical analysis of soils. *Soil Sci.*, **23**:343–353.

Casagrande, A. 1934. *Die Araometer-Methode zur Bestimmung der Kornverteilung von Boden und anderen Materialien.* Julius Springer, Berlin.

Day, Paul R. 1950. Physical basis of particle size analysis by the hydrometer method. *Soil. Sci.*, **70**:363–374.

Day, Paul R. 1953. Experimental confirmation of hydrometer theory. *Soil Sci.*, **74**:181–186.

Hough, B. K. 1957. *Basic Soils Engineering*. Ronald Press, New York.
Grim, R. E. 1953. *Clay Mineralogy*. McGraw-Hill Book Co., New York.
U.S. Bureau of Reclamation. 1960. *Design of Small Dams*.

PROBLEMS

(Refer to table below.)

1. What percentage by weight would you classify as fine sand?
2. What percentage by weight consists of particles finer than sand?
3. What percentage by weight would you classify as gravel?
4. Plot the analysis as a grain-size distribution curve.

   Wt. Original sample 384 g
   Wt. after prewashing 368 g
   Washing loss 16 g

Sieve	*Sieve Opening*, mm	*Wt. Retained on Sieve*, g	*Passing Sieve* Weight, g	%
2″	50.8	0	384	100.0
1½″	38.1	11	373	97.1
1″	25.4	65	308	80.2
¾″	19.1	58	250	65.1
¼″	6.35			
No. 4	4.75	63	187	48.7
No. 10	2.00	5	182	47.4
No. 40	0.42	18	164	42.7
No. 60	0.25			
No. 100	0.149	58	106	27.6
No. 200	0.074	89 − 3 = 86	20	5.2

Error = 3 g = 0.8%

5. What is the uniformity coefficient?
6. What is the coefficient of gradation?
7. A soil analysis shows it to contain 50% silt, 45%, and 5% sand. What is the texture? (Refer to Figure 4-7.)
8. A soil analysis shows it to contain 60% sand, 30% silt, and 10% clay. What is its texture?
9. Two hundred cc of soil is found to weigh 350 g. It contains 25% moisture by weight.
   (a) What is its bulk density?
   (b) What is its porosity?
   (c) What is its voids ratio?

Chapter 5 DETERMINING THE NATURE AND EXTENT OF A SUBSURFACE DRAINAGE PROBLEM

The first step in the solution of any drainage problem is to obtain information about the water tables, the soils, the topography, and other factors that will influence the design of a satisfactory drainage system. In this chapter we shall discuss some of the factors considered to be important to the design of a drainage system.

THE WATER TABLE

The water table is the upper limit of the waterlogged soil. It can be determined by digging a hole in the soil and observing the height to which the hole fills with water. In the field the height of the water table is commonly observed by means of observation wells. These are open holes which are partially filled with gravel, and a perforated pipe is placed in the hole. The region around the pipe is then backfilled with gravel so that water can flow freely into and out of the pipe and the hole. The observation well will give accurate measurements of the height of the water table, provided no upward pressure is produced by water flowing from confined aquifers at some distance below the soil surface.

Often when a hole is drilled in stratified soil the upper and denser layers of clay texture may appear to be dry. When the auger strikes a more permeable, sandy soil, however, the water rushes into the hole giving the appearance of an artesian pressure layer. But the assumption of artesian pressures is often wrong, for the water moves quickly into the hole because the sandy layer has higher permeability than the clay layers above it. Clay soils may be saturated, even though they do not appear to be saturated.

The simplest form of an observation well is an auger hole left in the field for casual measurement. The most valuable measurements can be obtained if an automatic water-stage recorder is placed on an observation well to give continuous record of water-table fluctuations. A careful interpretation of these water-table fluctuations can often lead to an accurate analysis of the source of the drainage water without any further measurements.

In the investigation of a regional drainage problem all existing data on ground-water levels should be accumulated and analyzed. Information on the ground water can be plotted in several different ways. The actual contours of the ground-water levels can be plotted on a contour map. A more valuable

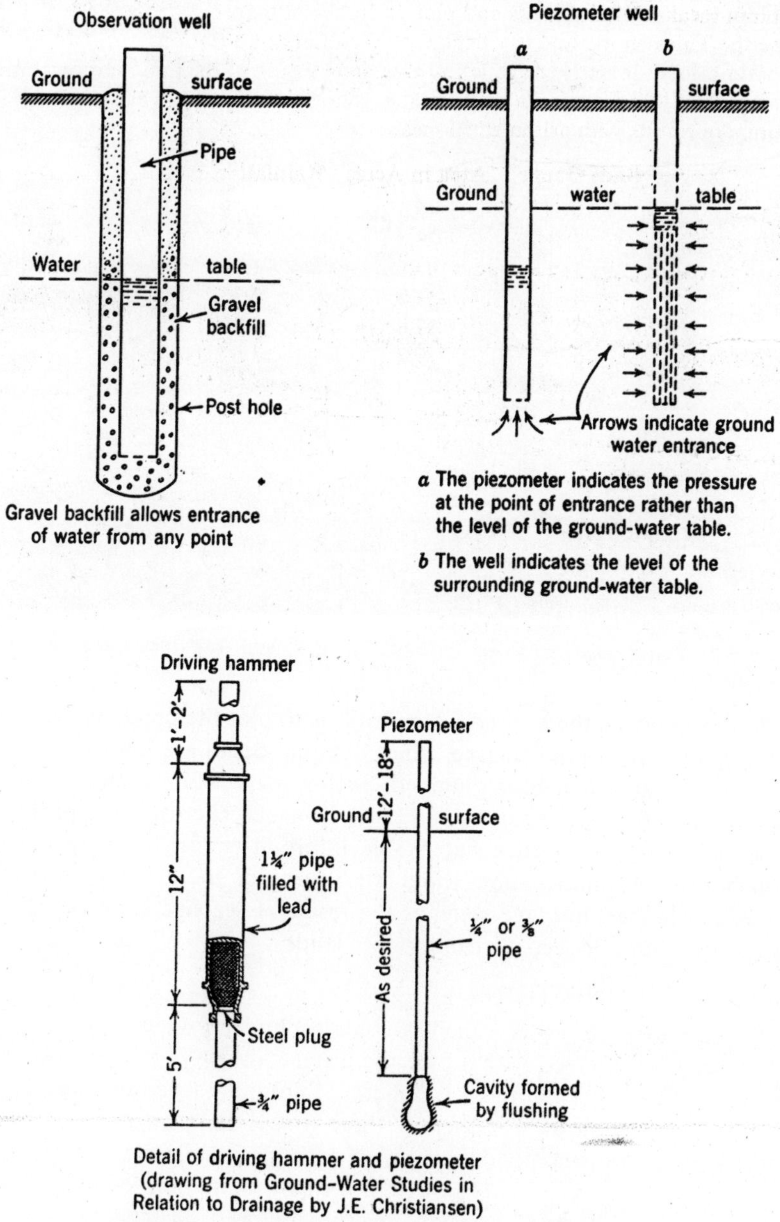

Figure 5-1 Details of observation wells and piezometers.

Figure 5-2 Water-stage recorder installed in the field for measuring water-table fluctuations.

method of plotting the ground-water levels is to plot the depth of the ground water below the ground surface. Lines of equal depth are joined and a map similar to a contour map is obtained. Such a map clearly indicates areas of high water table and indicates the areas in which the drainage problem is most acute. A plot of the water-table information on the map can also furnish valuable information about the source of the drainage water; for example, whether drainage water is the result of seepage from rivers, from excess irrigation, or seepage from higher lands.

ARTESIAN CONDITIONS

Artesian water is water which is confined under pressure in an aquifer. If the aquifer is pierced with a pipe the water will rise in the pipe above the ground surface. Artesian water is difficult to drain because the drains have to be placed much closer together in these areas. It is important, then, to detect artesian pressure conditions before a drainage system is put into operation. Artesian pressures can be determined by examining existing artesian wells, by observation of artesian pressure by well drillers, or any other information that is locally available. If artesian pressures are suspected and no definite information is available, a series of piezometers can be placed in the ground. A piezometer is an open-ended pipe which is placed beneath the soil surface. The purpose of this pipe is to measure the soil-water pressure at the end of the pipe. The water pressure is measured by the elevation of the water in the pipe. The pressure head is given by the distance between the end of the pipe and the elevation of the water in the pipe itself. Piezometers are usually placed in batteries, that is, several piezometers are placed side by side, each at a different depth. The purpose of the piezometer is to measure the vertical gradients of water in the soil to see whether or not there is any artesian pressure. Artesian pressure is indicated if the water in the deeper piezometers rises to a greater elevation than the water in the shallower piezometers. Small-diameter pipe is used for piezometers and may be from $\frac{3}{8}$ to $\frac{1}{2}$ inch in diameter. It is possible to jet these piezometers to almost any depth up to several hundred feet with the proper equipment. A jetting rig that furnishes pressures of 300 to 400 pounds per square inch and a volume of water of 10 to 20 gallons per minute is frequently adequate for jetting these small bore pipes. In the event that a gravel stratum is reached and must be penetrated by the piezometer, it is necessary to use driller's mud or bentonite clay to seal off the gravel layer and prevent the loss of the boil.

SOILS INFORMATION

Detailed information concerning the soils in the area to be drained is basic to any drainage design. In the design of a system on an individual farm, it may be possible for the designer to obtain the necessary information by boring a series of holes. On the other hand the information may be obtained from the boring of the observation wells or from the jetting of piezometers.

On the larger projects it is essential to have a complete soil survey made of the area. For reconnaissance purposes a scale of 1:50,000 is probably adequate. However, for design purposes one should have maps on a scale of about 1:5,000. Aerial photographs are most suitable for the purpose of soil surveying.

On a soil survey, the soil borings are made to a depth of about three to six feet. This is adequate in most humid regions. For the purpose of drainage design in irrigated areas it is necessary to have deeper borings. The depth of the deeper borings will depend on the area. It may be necessary to probe to depths greater than 100 feet.

The following information should be indicated on the soils maps.

Soil Texture as Determined in the Field by the Soils Technician

Laboratory checks should be made of the observations made in the field. It is often possible for the texture to be evaluated incorrectly in areas where the soil contains salt or in areas of lateritic soils containing iron inclusions.

Depth to the Impermeable Layer

The depth to the impermeable layer is important in the location of drains. Most of the drainage formulas for determining the depth and spacing of the drains require information about the depth of the impermeable layer.

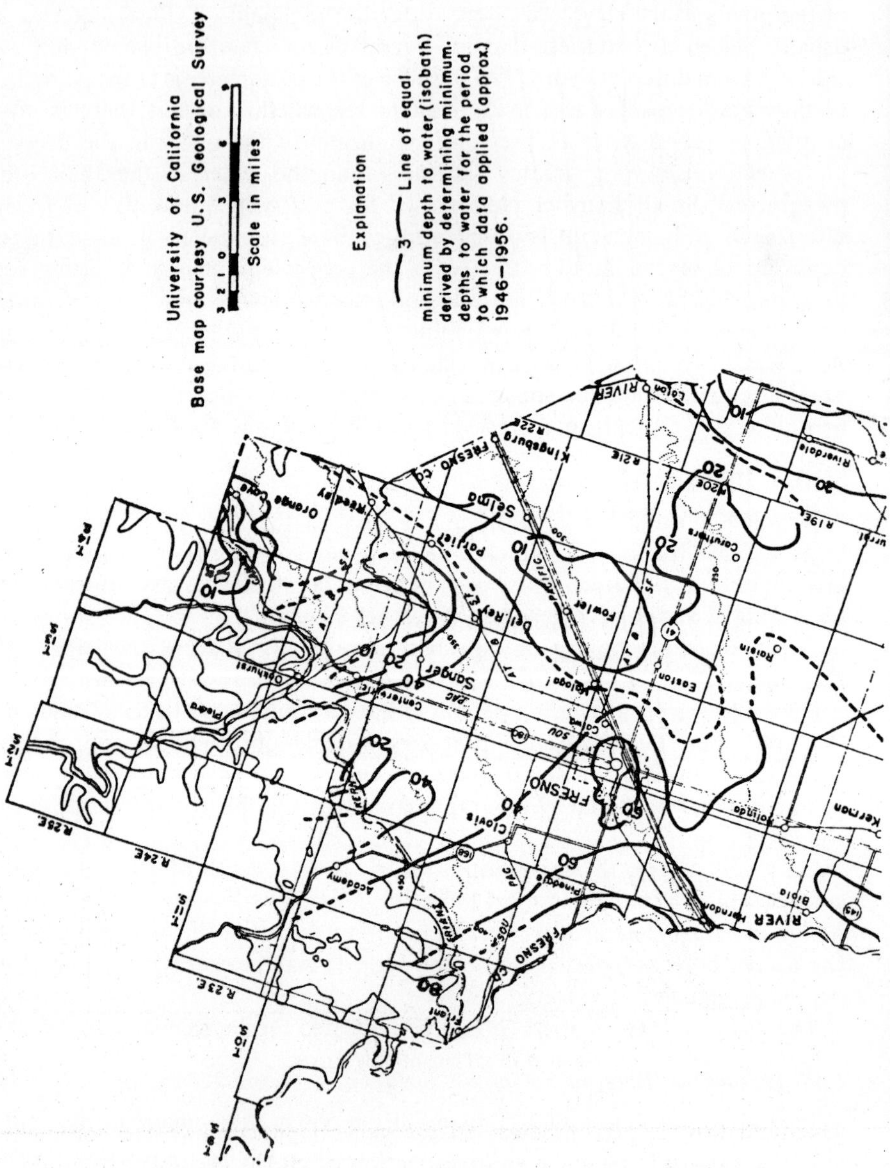
University of California
Base map courtesy U.S. Geological Survey
3 2 1 0 3 6 9
Scale in miles
Explanation
— 3 — Line of equal minimum depth to water (isobath) derived by determining minimum. depths to water for the period to which data applied (approx) 1946—1956.
FRESNO
Clovis
Sanger
Selma
Reedley
Orange Cove
Kingsburg
Fowler
Del Rey
Parlier
Easton
Caruthers
Riverdale
Raisin
Kerman
Biola
Academy
Centerville
Piedra
Pinedale
Herndon
FRESNO RIVER
R.25E.
R.24E.
R.23E.
R.22E
R.21E
R.20E
R.19E

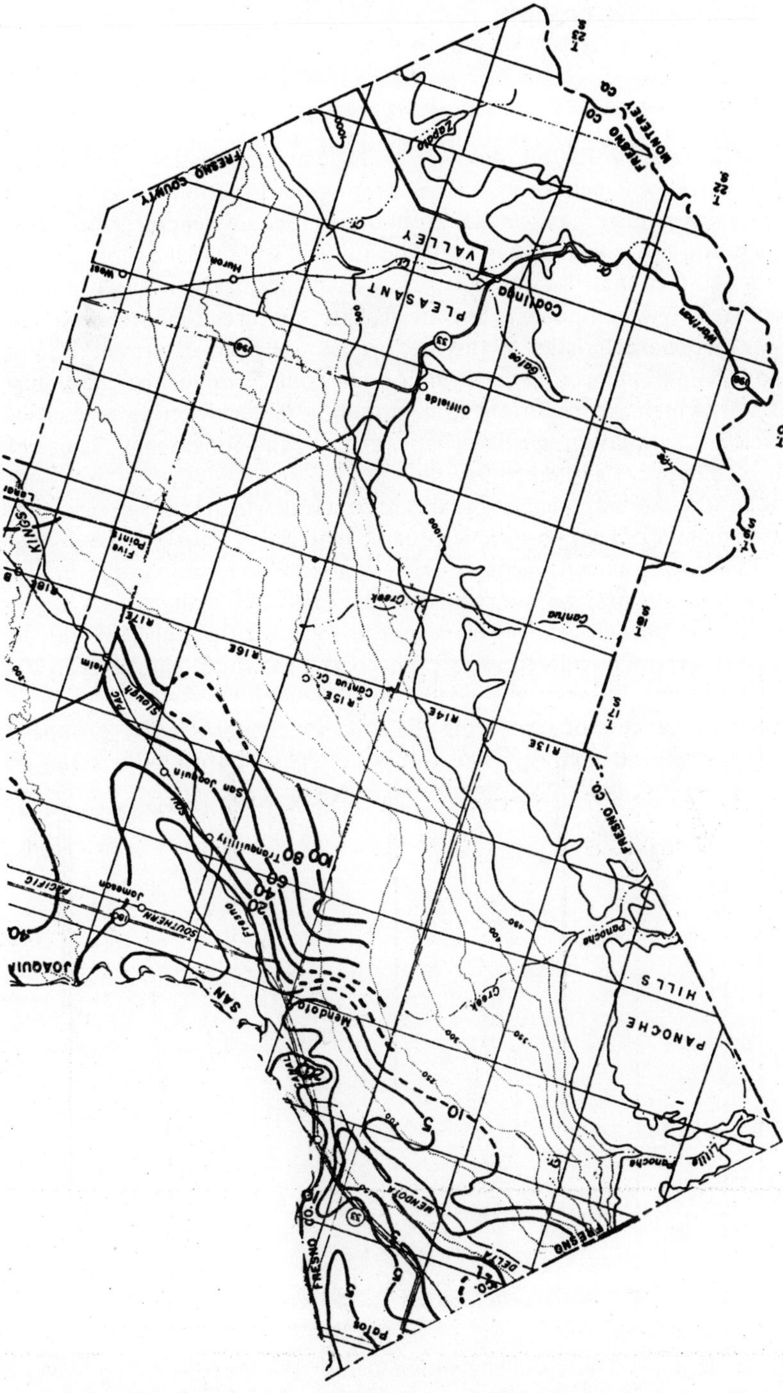

Figure 5-3 Fresno County. Lines of equal minimum depth to water during period approximately 1946–1956.

Figure 5-4 Installing piezometers by jetting.

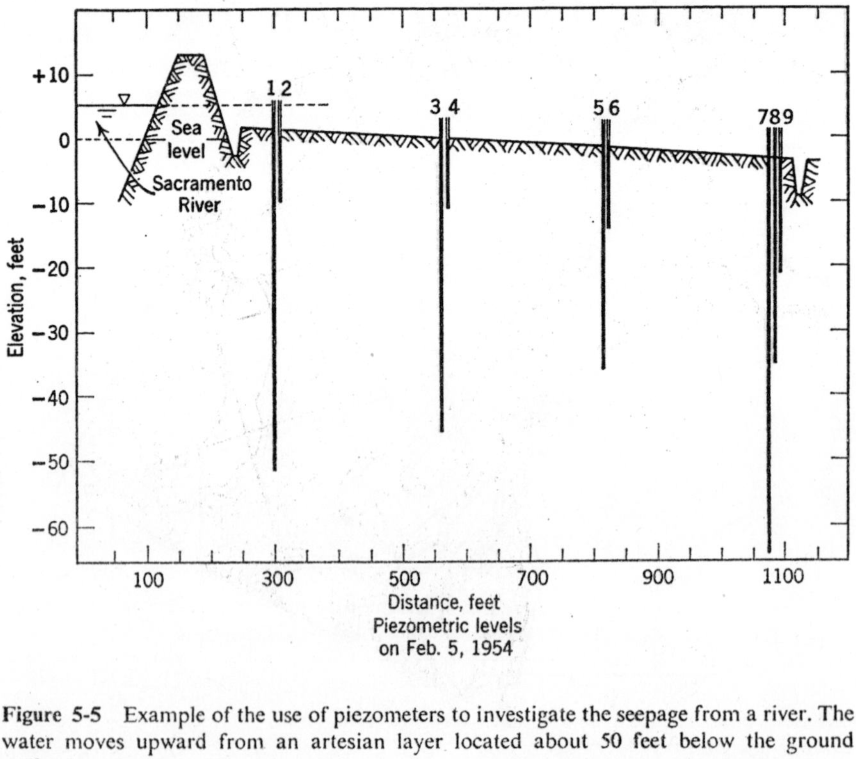

Figure 5-5 Example of the use of piezometers to investigate the seepage from a river. The water moves upward from an artesian layer located about 50 feet below the ground surface.

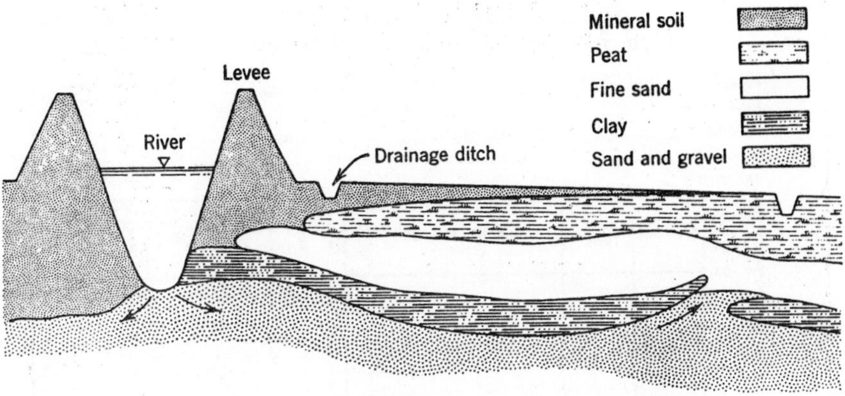

Figure 5-6 Typical section of underground strata.

Figure 5-7 Rotary drill used to investigate soil conditions.

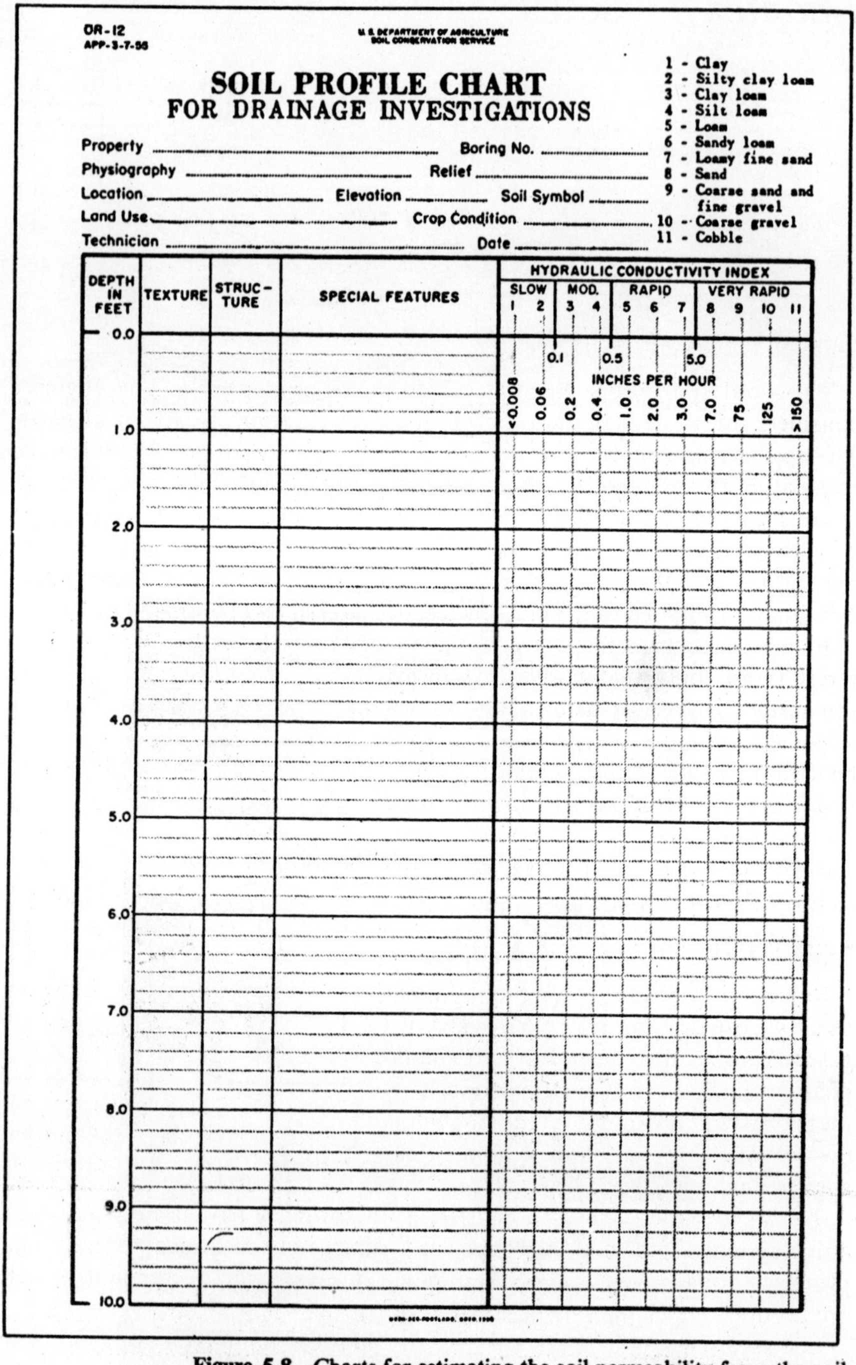

DR-12
APP-3-7-56

U.S. DEPARTMENT OF AGRICULTURE
SOIL CONSERVATION SERVICE

SOIL PROFILE CHART
FOR DRAINAGE INVESTIGATIONS

1 - Clay
2 - Silty clay loam
3 - Clay loam
4 - Silt loam
5 - Loam
6 - Sandy loam
7 - Loamy fine sand
8 - Sand
9 - Coarse sand and fine gravel
10 - Coarse gravel
11 - Cobble

Property .... Boring No. ....
Physiography .... Relief ....
Location .... Elevation .... Soil Symbol ....
Land Use .... Crop Condition ....
Technician .... Date ....

DEPTH IN FEET	TEXTURE	STRUCTURE	SPECIAL FEATURES	HYDRAULIC CONDUCTIVITY INDEX
				SLOW 1 2 · MOD. 3 4 · RAPID 5 6 7 · VERY RAPID 8 9 10 11
				0.1 · 0.5 · 5.0
				INCHES PER HOUR: <0.008 · 0.06 · 0.2 · 0.4 · 1.0 · 2.0 · 3.0 · 7.0 · 75 · 125 · >150
0.0				
1.0				
2.0				
3.0				
4.0				
5.0				
6.0				
7.0				
8.0				
9.0				
10.0				

Figure 5-8 Charts for estimating the soil permeability from the soil

It is a generally accepted practice to classify soils with respect to texture and then to adjust the hydraulic conductivity, "HC" estimates with respect to soil structure, alkali and other influencing factors. Soil texture refers to the relative proportions of the various size groups of individual soil grains in a mass of soil. Structure refers to the condition of the soil grains, (clay, silt, sand, etc.), in the way they are arranged and bound together into aggregates with definite shape.

Texture Legend		Structure Legend
Clay ------------- c	Loamy fine sand --- lfs	Massive ------------ m
Silty clay loam -- sicl	Sand -------------- s	Platy -------------- pl
Clay loam -------- cl	Coarse sand ------- cos	Prisimlike --------- pr
Silt loam -------- sil	Fine gravel ------- fg	Blocky ------------- bk
Loam ------------- l	Coarse gravel ----- cg	Granular ----------- gr
Sandy loam ------- sl	Cobble ------------ co	Single grain ------- sg

Aggregate length and thickness has an effect upon the "HC". The overlap of aggregates having a horiz. axes 3 or 4 times longer than the vertical can have a marked effect upon "HC". The "HC" of stratum can be changed by cracks, crevices and fractures. Some fractured, (blocky structured), clay or shale stratum are often much higher than sands or gravel stratum. The "HC" of sandy stratum are generally higher when the grains are round and about the same size than when they are irregular in shape and of different sizes. Flat grains tend to overlap and reduce "HC" rates. The matric rather than the coarser material in a gravel or cobble stratum govern the "HC" and should be used as a basis in estimating the "HC".

The following two tables are suggested as a possible guide in rating the "HC" of various stratum during a drainage soil survey. It is not anticipated that these tables will replace laboratory analysis and actual field "HC" measurements but that they will be used as estimates and bolstered or revised as additional experience and data becomes available.

HYDRAULIC CONDUCTIVITY INDEX NUMBER RANGE for TEXTURE - STRUCTURE CORRELATION

STRUCTURE / TEXTURE	MASSIVE	PLATY	PRISMLIKE	BLOCKY	GRANULAR	SINGLE GRAIN
	GENERAL INDEX NUMBER RANGE					
CLAY	1	1-4	1-2	1-8	2-7	
SILTY CLAY LOAM	2	1-4	1-4	2-7	3-7	
CLAY LOAM	3	2-3	2-3	3-7	3-7	
SILT LOAM	4	3-4	2-4	3-6	4-6	
LOAM	5	3-5	4-5	5-7	5-7	
SANDY LOAM		4-6	5-6	6-7	6-8	6
LOAMY FINE SAND				6-7	7-8	7
SAND						8
COARSE SAND						9

ESTIMATE of ALKALI INFLUENCE on HYDRAULIC CONDUCTIVITY by pH COLOR INDICATORS

SOIL TEXTURE	pH ESTIMATES FROM COLOR INDICATORS							
	8.6		9.0		9.6		10.0	
	HC REDUCED %	INDEX NO.	HC REDUCED %	INDEX NO.	HC REDUCED %	INDEX NO.	HC REDUCED %	INDEX NO.
CLAY	20	1	40	-1	70	-1	90	-1
SILTY CLAY LOAM	15	2	30	2	60	1	80	1
CLAY LOAM	15	3	25	3	50	2	70	2
SILT LOAM	15	4	25	4	50	3	70	2
LOAM	10	5	20	5	40	4	50	4
SANDY LOAM	5	6	15	6	35	5	45	5
LOAMY FINE SAND	5	7	15	7	30	6	40	6
SAND	5	8	10	8	25	8	40	7

texture and the soil structure. Courtesy of the Soil Conservation Service.

The impermeable layer may be a compressed soil that has been compressed by the weight of the glaciers during the ice age. Such soils are found extensively in the north central states. In other more southerly regions there is a dense subsoil of clay texture. Soils of this nature are classified as planosols and are found extensively in southern Iowa, northern Missouri, and in parts of Illinois and Ohio as well as other states.

The more recent alluvial soils of the irrigated west have impermeable layers because of the stratification of the sediments that have been deposited from the streams.

The term "impermeable" is a relative term. All soils are more or less permeable. Deep seepage occurs, to a limited extent, even on very slowly permeable subsoils. However, if the permeability of the subsoil is about one-tenth that of the surface soil, it can be considered impermeable from a drainage standpoint. There will be waterlogging above this layer if the rainfall rate, or the rate at which water is added to the soil, exceeds the permeability of this layer. The flow pattern of the water moving towards the drains will be altered drastically by this layer of lower permeability. The drains will have to be placed close together to achieve the same effect as in a deep permeable soil.

Waterlogged Conditions

The presence or absence of a water table, and the depth to the water table at the time of the survey, should be indicated. Sometimes the soil coloring is indicative of waterlogging. An orange and black mottling of the soil is due to ferric and manganic ions, and indicates periodic reducing and oxidizing conditions. The depth and extent of this mottling should be noted since it may provide the data required for the location of the drain. The presence of mottling in the soil may indicate soil conditions that have caused waterlogging in the past.

The occurrence of blue clay or blue soil will indicate reducing conditions because of continuous submergence. Under these conditions the iron in the soil comes into solution as ferrous iron, which imparts the blue color to the soil.

OUTFLOW FROM DRAINAGE SYSTEMS

Wherever possible, data should be obtained on the outflow from existing drainage systems. These data are then used to determine a unit-area discharge for drainage in a particular area. It should also be observed whether or not the drainage systems from which the data are obtained are adequate for the area. If they are not adequate a closer spacing of drains would result in a larger outflow from the drainage system. Wherever possible the data obtained as outflow from drainage systems should be correlated not only with the soil type, but with other characteristics such as texture and permeability. By accumulating data of this kind it is possible to extrapolate data obtained in one area for use in other areas. The compilation of these data should be done by an engineer in conjunction with a soils expert who is conversant with the soil conditions in the area.

The drawdown, capacity, and area of influence of pumped wells should be determined wherever data are available. The possibility of using wells for drainage should not be overlooked.

TOPOGRAPHY

Good topographic maps are essential to a study of a drainage problem since they frequently provide the key to the essential features of the system, such as the location and type of outlet (gravity or pumped), surface slopes, and the degree of land preparation. The accuracy of the vertical control is particularly important. For preliminary studies a vertical interval of two to five feet may be adequate. Detailed studies require an interval of one foot or less in flat land. The topographic survey should indicate the position, alignment, and gradient of existing ditches, streams, culverts and other natural and artificial features that may influence the drainage system. Possible outlets should also be included. Where the data are available, the capacities and channel dimensions of both the drainage channels and the natural water courses should be determined. In addition, any data available on runoff should be collected, and information concerning the stability of channel side slopes should also be gathered. The location of pumped wells should also be indicated along with their capacities, drawdowns, and zones of influence. The use of aerial photos in conjunction with a ground survey frequently will be of assistance in locating areas of poor drainage and in establishing cultural features.

REFERENCES

Blaney, H. F. and C. A. Taylor. 1931. Soil sampling with a compressed air unit. *Soil Sci.*, **31**(1).

Donnan, W. W. and G. B. Bradshaw. 1952. Drainage investigation methods for irrigated areas in western United States. *U.S. Dept. Agr. Tech. Bul.* 1065.

Donnan, W. W., G. B. Bradshaw, and H. F. Blaney. 1954. Drainage investigation in Imperial Valley, California, 1941–1951 (A Ten-Year Summary). SCS TP-120, p. 16.

Donnan, W. W. and J. E. Christiansen. 1944. Piezometers for ground water investigation. *Western Construct. News*, **19**:77–79.

Christiansen, J. E. 1943. Ground water studies in relation to drainage. *Agr. Engr.*, **24**:339–342.

Israelsen, O. W. and W. W. MacLaughlin. 1935. Drainage of land overlying an artesian ground-water reservoir. *Utah Agr. Exp. Sta. Bull.* 242 (progress report) 1932 and *Bull.* 259 (final report) 1935.

Meinzer, O. E. 1927. Method of exploring and repairing leaky artesian wells. *U.S. Geol. Serv. W.S.P.* 596-A.

Pillsbury, A. F. and J. E. Christiansen, 1947. Installing ground water piezometers by jetting for drainage investigations in Coachella Valley, *Calif. Agr. Eng.*, **28**:409–410.

Reeve, R. C. and Max. C. Jensen. 1949. Piezometers for ground-water flow studies and measurement of subsoil permeability. *Agr. Eng.*, **30**:435–438.
Reger, J. S., A. F. Pillsbury, R. C. Reeve, and R. K. Peterson. 1950. Techniques for drainage investigations in Coachella Valley, *Calif. Agr. Eng.*, **31**:559–564.
Roe, H. B. and Q. C. Ayres. 1954. *Engineering for Agricultural Drainage*. McGraw-Hill Book Co., New York.
Schlick, W. J. 1918. Spacing and depth of laterals in Iowa under drainage systems and rate of runoff from them. *Iowa Experiment Sta. Bull.* 52.
Tolman, C. F. 1937. *Ground Water*, 1st edition. McGraw-Hill Book Co., New York.
Uhland, R. E. and A. M. O'Neal. 1951. Soil permeability determinations for use in soil and water conservation. U.S.D.A. SCS TP-101.

Chapter 6 STATICS OF SOIL WATER

THE NATURE OF WATER

Electrical Properties

Water consists of atoms of oxygen and of hydrogen. The arrangement of these atoms is such that the water molecule has a localized distribution of electrical charges at its outer extremities. Because of the separation of the positive charge from the negative charge this type of molecule is called a *dipole*. Dipoles can be attracted by electrical forces, a property of water which is used in electro-osmosis.

Density

The arrangement of charges on the water molecule causes it to have an open structure and a low density of 1 g/cc at 4°C. At 20°C the density is 0.99823 g/cc. For practical purposes it is assumed that water has a constant density of 1 g/cc.

FORCES HOLDING WATER IN SOILS

Imbibitional forces

In the previous section we noted that the molecules of water possess an electrical charge and are dipoles. In effect each water molecule acts as though it were a small bar magnet.

The clay particles in the soil also possess a net electrical charge and can attract the water molecules. The effectiveness of the attractive forces diminishes rapidly with the distance from the clay particles hence this adsorbed layer of water molecules may be rather thin, probably 3 to 10 molecules thick or 8 to 28 Å.

Some clays, such as the montmorillonitic clays, may absorb water into the crystal lattice of the clay itself. This water, called interplanar water, causes the swelling and shrinkage of the clay.

Since clays possess a net negative electrical charge they attract cations to their surface. Some cations such as sodium are surrounded by a shell of water molecules called water of hydration.

A comparison between the crystal lattice radii of various ions and their hydration radii in angstroms is given in the table below.

Ion	Na	K	NH_4	Rb	Cs
Crystal Lattice Radii, Å	0.98	1.33	1.43	1.49	1.63
Hydration Radii, Å	7.0	5.3	5.4	5.1	5.0

All of the above forms of water are grouped under the general term of imbibitional water. The imbibitional forces which bind the water to the soil solids are very strong. Imbibitional water is not considered available for plant growth nor does it enter into the scheme of soil-water movement. In drainage work imbibitional water is relatively unimportant.

Capillary Forces

The rise of water in tubes of capillary dimensions is due to a difference in the attractive forces of water to water and water to glass (if the capillary is made of glass). The water-to-water forces are called *cohesive* forces and the water-to-solid forces are *adhesive forces.*

If the adhesive forces are greater than the cohesive forces the liquid will rise on the surface of the solid until the surface of the water at its intersection with the glass is perpendicular to the resultant of the two forces. If the ad-

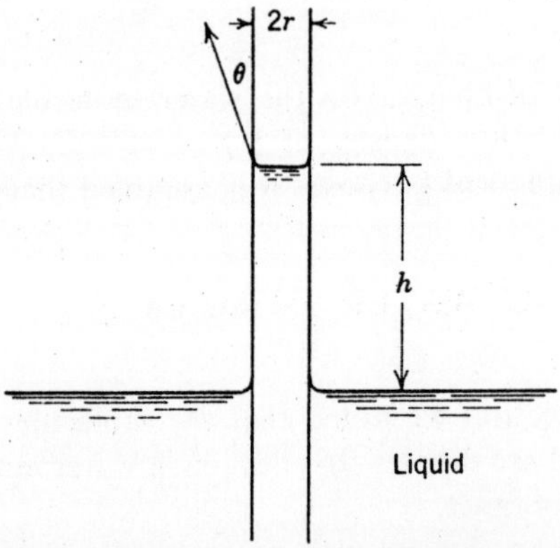

Figure 6-1 Rise of water in tubes of capillary dimensions.

hesive force is very much greater than the cohesive force the contact angle will be zero. This is true for water and glass. It is also generally considered to be so of water in soils. Some research work has shown that the adsorbed cations on the surface of the clay particles have an effect on the contact angles. For our purposes the rather simple description given above will suffice.

If we now consider a tube of capillary dimensions inserted into water we note that water will cling to the walls of the tube with a force of $2\pi r\sigma \cos \theta$, where $2\pi r$ represents the perimeter of the capillary and σ is the surface tension of the water. The angle of contact is θ, and $\cos \theta$ resolves the force in the vertical direction.

The vertical-force component supports the weight of the fluid column of height h. If ρ is the density of the fluid we can equate the downward force of the weight of water suspended in the capillary to the upward force of surface tension.

$$2\pi r\sigma \cos\theta = g\rho\pi r^2 h \tag{1}$$

and

$$h = \frac{2\sigma \cos\theta}{g\rho r} \tag{2}$$

If the contact angle is zero, $\cos\theta = 1$ and equation 2 becomes

$$h = \frac{2\sigma}{g\rho r} \tag{3}$$

Other Forces

In addition to the forces discussed above there may be other forces resulting from chemical and physical actions. These other forces are relatively minor and usually can be disregarded in drainage design.

The Energy Concept of Soil Moisture

Potential energy is the latent or potential ability to do work. It may result from the position of the water with regard to some reference datum, or it may be caused by pressure that results either from an externally applied force or from the action of gravity.

In the region above the water table the soil-moisture pressure is negative. The soil exerts an attraction on the soil water. This attraction of the soil for the water can be considered equivalent to work performed on the water. The standard of reference for the amount of work is taken to be a free water surface. The water molecules at the surface of a body of water are considered to have no potential energy, whereas the water that is held by the soil possesses a negative amount of energy because of the work done on it by the soil. The amount of work, which is a potential function, is defined as the amount of work that is required to move a unit mass of water from a point in the soil to a free water surface.

The significance of the work function or of the potential of the soil water can be seen in Figure 6-2. In the figure a homogeneous column of soil is allowed to come into isothermal equilibrium with a free water surface at its lower end. Since the system is in equilibrium there must be a balance of forces on each element of water in the column. If we consider a small volume of water at point A we can see that there will be an upward force because of capillarity. The downward force of gravity will oppose the upward capillary force. The work required to move the water to point A will be equal to the work done by the capillary forces over a distance h. This work will be accomplished in opposition to the force of the gravitational field. Since work is equal to force multiplied by distance, the potential of the water at A is equal to the distance h multiplied by the force due to gravity, or is equal to hgm, where m is the mass of the water and g is the acceleration due to gravity.

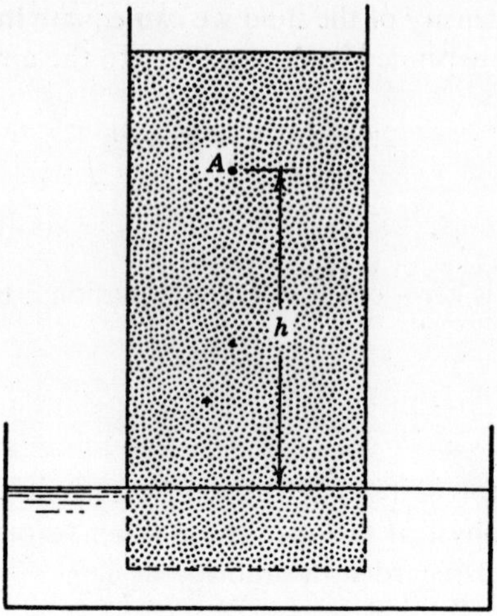

Figure 6-2 Diagram of soil column in moisture equilibrium with a free-water surface (after Russell).

SOIL-MOISTURE PRESSURES

The Water Table

The water table is defined as the locus of points at atmospheric pressure. Below the water table, soil-water pressures are positive. The water table represents the upper limit of the zone of positive pressures. Above the water

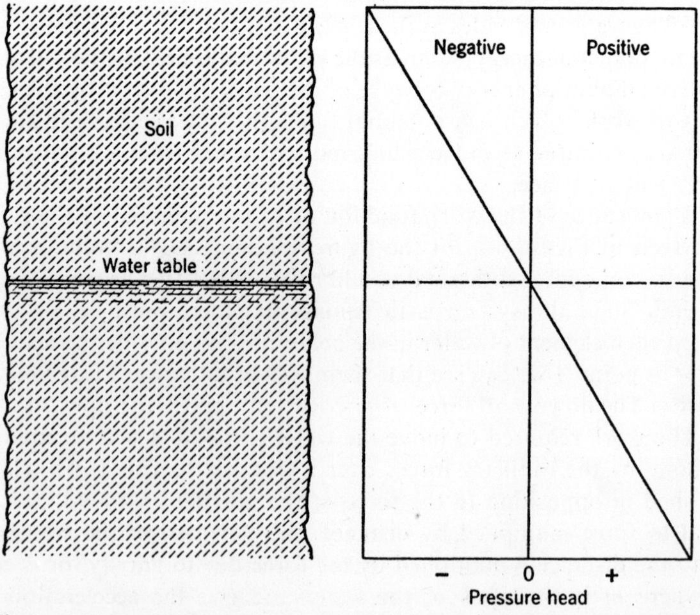

Figure 6-3 Equilibrium-pressure distribution in a soil with a water table.

table the pressures of the soil water are less than atmospheric and the water pressure is negative.

The important thing is that the water phase is continuous as the pressure changes from positive to negative in much the same way that water is held in a capillary tube that has one end in a pail of water.

NEGATIVE SOIL-MOISTURE PRESSURES—TENSION, SUCTION, CAPILLARY PRESSURE

Above the water table the soil-moisture pressure is negative. Various terms have been used to describe this pressure condition. In general the terms are all synonymous. Soil-moisture suction is currently favored by most of the soil scientists. Others favor the use of the word "tension." The petroleum engineers use the term capillary pressure, and it is this term that will be used interchangeably with negative soil-moisture pressure. It should be noted that the capillary pressure is the negative of the soil-moisture pressure.

One more word about the use of the term capillary pressures. In drainage work the negative pressures above the water table are relatively small in magnitude and are due to capillary forces. However, in soils on which plants are growing these negative pressures may be on the order of several atmospheres. At these large values of negative pressure the capillary forces may not be the dominant forces.

If a pad of soil is placed on a porous plate and a negative pressure is applied to the porous plate, water will be extracted from the soil. The more negative pressure applied, the more water will be removed.

Similarly if we apply positive *air* pressure to the *soil*, water will be forced out of the soil through the porous plate.

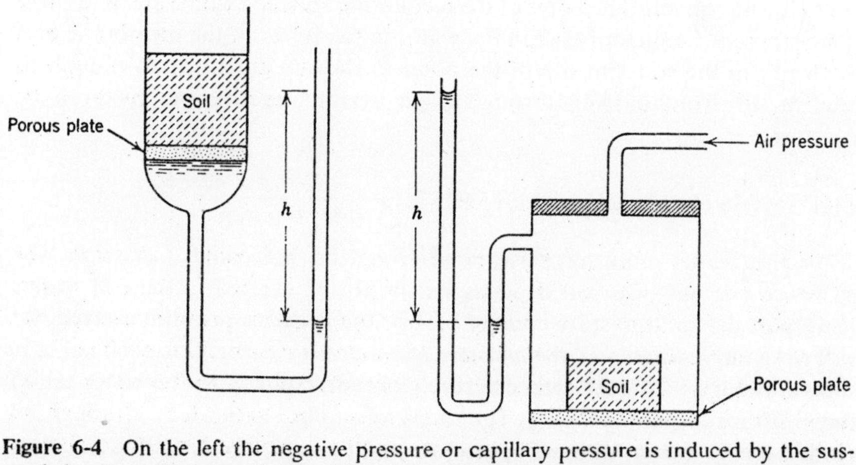

Figure 6-4 On the left the negative pressure or capillary pressure is induced by the suspended column of water. On the right the capillary pressure is induced by a positive air pressure in the pressure-membrane apparatus. If the pressure is the same in both instances (the positive air pressure is equal in magnitude to the negative water pressure) then the moisture content of the soil will also be the same.

In each case the forces holding the water in the soil are the result of surface tension. The forces depend on an air-water interface. If the pressure is changed on either side of the interface, the interface will move to a new position where equilibrium of the forces is established once again.

In each case the porous plate is saturated with water. The pores of the plate are sufficiently small so that they will not drain at the pressures used.

In case I a negative pressure is applied to the soil water through the porous plate. The pressure of the air in the soil remains one atmosphere.

In case II a positive pressure is applied to the air inside the bell jar. The air pressure at the lower side of the porous plate remains one atmosphere.

If the negative water pressure in case I is exactly the same as the positive air pressure in case II the soil will be drained to the same moisture content in each case.

Case I (negative water pressure) can be used to apply pressures of about $-\frac{2}{3}$ atmosphere or -20 feet ($-6\frac{2}{3}$ meters). Below this value (more negative) the water tends to separate in the column.

On the other hand case II (positive air pressure) can be used to obtain very high values of air pressure, up to 160 amospheres, which correspond to very low values of water pressure (-160 atmospheres). Since this is far below the range of moisture available to plants, the method is of great importance in studying the relationship among plants, soil, and water. Richards (1949) has utilized this method in his pressure plate and pressure-membrane apparatus. Both of these pieces of apparatus utilize the same basic principle, but the difference in name refers to the porous material used to separate the soil from the atmosphere.

Negative soil-moisture pressures can be measured directly by a device called a tensiometer. A tensiometer consists of a porous membrane in close contact with the soil. The pores of the membrane are filled with water and there is a continuous contact between the water in the pores of the membrane and the water in the soil.The size of the pores in the cup are all small enough to prevent air from passing through them at the negative pressures to be measured.

THE MOISTURE CHARACTERISTIC

As the reader may have surmised from the preceding discussion, the moisture content of a soil depends on the forces exerted in the soil water. If we plot the soil-moisture content against the negative pressure exerted, we will get a curve known as the *moisture characteristic curve*. One such curve is shown in Figure 6-5. At zero pressure (this corresponds to the water table) the moisture content is 47.1%. The soil is said to be "saturated" although, in reality, it is extremely difficult to saturate a soil completely. Usually soil below a water table is only about 90% saturated because of the air that is trapped in the soil pores.

As the pressure changes from zero to -20 cm there is very little change in

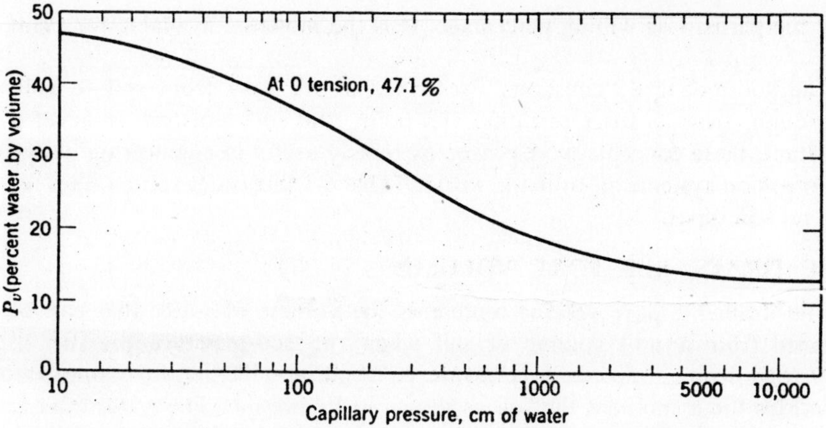

Figure 6-5 Moisture characteristic for Ida silt loam, bulk density 1.31 gm/cm³.

the moisture content. This means that there are no pores large enough to be drained in this range of pressures.

Decreasing the pressure head below −20 cm causes a rapid decrease in moisture content, because most of the soil pores are of a size that is drained by these negative pressures.

SOME SOIL-MOISTURE "CONSTANTS"

Soil moisture is rarely, if ever, constant at any pressure. It is always being subjected to pressure gradients and vapor-pressure differences that cause it to move.

However, it has been found experimentally that certain moisture contents are of particular significance in agriculture. These moisture contents vary from soil to soil and are best identified by a particular soil-water pressure.

The *field capacity* of a soil has been defined by Veihmeyer and Hendrickson as the moisture content of a soil after downward movement has materially decreased. It is the amount of water retained in the soil after an irrigation, say about three days after, or after a rain. The soil-water pressure at field capacity varies from soil to soil, but tests have shown that $-\frac{1}{10}$ to $-\frac{1}{3}$ atmosphere is the range.

The pressure-plate apparatus can be used to determine the field capacity, or the capacity can be determined in the field by sampling the soil. A relatively easy laboratory determination can be made with the centrifuge. A small sample of soil is whirled in a centrifuge with a centrifugal force of 1000 gravity. The moisture remaining in the sample is called the *moisture equivalent* and, as the title suggests, is about the same as the field capacity.

The *permanent wilting percentage* is the amount of water left in the soil when the plant is unable to extract any more. The soil-moisture pressure is equal to about −15 atmospheres at the P.W.P.

The *available moisture* represents the difference between the field capacity and the permanent wilting percentage. It is the moisture available for plant use.

The soil-moisture "constants" described above vary from soil to soil. Although, from a strict scientific viewpoint, the soil moisture is never a constant, these concepts have proven extremely useful in engineering design of irrigation systems all over the world. Table 6-1 illustrates some values for several soil types.

THE DRAINABLE PORE VOLUME

The drainable pore volume represents the volume of water that can be drained from a unit volume of soil when the soil-moisture pressure is decreased from atmospheric pressure to some specific negative pressure. Consider for a moment the soil as shown in Figure 6-6. The water table is

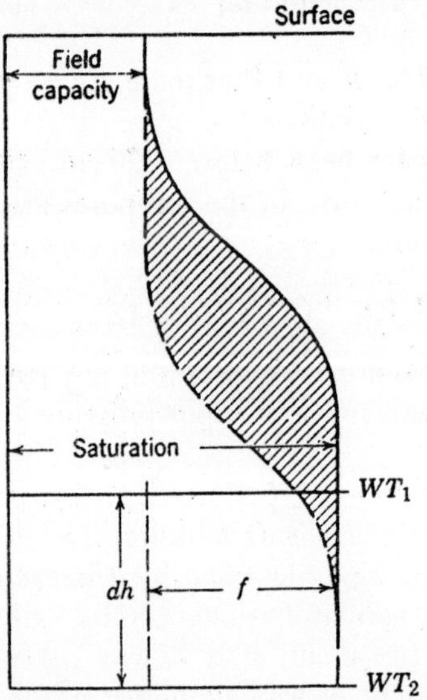

Figure 6-6 Hypothetical moisture profiles showing two stages of a falling water table in a deep soil. The shaded area represents the soil-water yield and is equal to the water which enters the drain. After Childs (1960).

shown at height h_1. Plotted on the same figure in curve 1 are the moisture contents of the soil above and below the water table. These moisture contents are equilibrium values. That is they will be obtained if the water remains stationary for sufficient time. For the rapidly moving water table the moisture distribution would be different. For a slowly moving water table the moisture contents will be a reasonable approximation of the values given in Figure 6-6.

Now assume that the water table drops to a new position, h_2. If we wait sufficiently long the moisture content will now be given by curve 2. The amount of water that has drained out of the soil must be equal to the decrease that has occurred in the soil moisture above the water table. Note that this decrease takes place an appreciable distance above the water table. That is, the water table does not represent a sharp demarcation line for the drainage of the soil.

Another way of expressing the changes that take place in the soil moisture when the water table drops is by replotting the soil-moisture characteristic curve. The moisture concentration is defined as the volume of water per unit volume of soil. The difference between the saturated moisture concentration and the moisture concentration at a given capillary pressure is the drainable pore volume for that pressure. It represents the change in moisture concentration when the soil-moisture capillary pressure changes from zero to the new value.

For example, suppose a soil contains 35% moisture by dry weight when it is saturated, and the soil-moisture pressure is equal to zero. When the soil-moisture capillary pressure is increased to 100 cm of head, the soil-moisture content becomes 30%. Since the bulk density of the soil is 1.40, 1 cc of dry soil

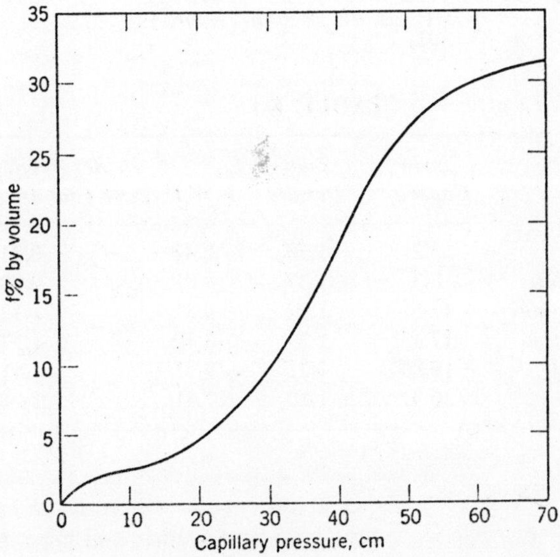

Figure 6-7 Drainable-porosity curve for a sandy loam soil.

will weigh 1.40 g. At saturation 1 cc of soil will contain $1.40 \times 0.35 = 0.49$ g of water. If we assume that water has a density of 1.00, the moisture concentration at saturation is 0.49 cc of water per cc of soil.

Similarly at a capillary pressure of 100 cm the soil-moisture concentration is $1.40 \times 0.30 = 0.42$ cc of water per cc of soil.

The drainable porosity at saturation is zero and at 100 cm is $0.49 - 0.42 = 0.07$ cc of water per cc of soil. This is the amount of water that will drain out of 1 cc of soil at the soil surface when the water table drops from the soil surface to a point 100 cm below the soil surface.

In the example just given the change in moisture content was calculated for a cubic centimeter of soil located at the soil surface. Usually it is desirable to know the quantity of water that drains out of a column of soil. This quantity is determined by adding up the quantities that drain out of each portion of the column. If we let f be the drainable porosity, then f is a function of the capillary pressure h and can be written $f(h)$. As the water table drops from h_1 to h_2, the quantity of water draining out of a unit column will be given by

$$q = \int_{h_2}^{h_1} f(h)\, dh \tag{4}$$

The function $f(h)$ is complicated for most soils, but most of the time it is possible to write an approximate expression for the relationship between the drainable porosity and the capillary pressure. The simplest equation to use is that of a straight line ah where a is the slope of the line. The quantity of water drained is now given by

$$q = \int_{h_2}^{h_1} ah\, dh = \frac{a}{2}(h_1^2 - h_2^2) \tag{5}$$

TABLE 6-1

Soil Type	*Field Capacity*	*Bulk Density*	*P.W.P.*	*Inches Available Water per Foot Depth of Soil*
Columbia sand	3.21	1.50	1.47	0.31
Hesperia sandy loam	8.3	1.37	4.05	0.70
Glenn fine sandy loam	12.5	1.32	4.5	1.32
Yolo loam	17.43	1.27	8.79	1.32
Ramona clay loam	19.80	1.25	9.75	1.51
Dublin clay adobe	30.97	1.19	17.61	1.91

REFERENCES

Baver, L. D. 1956. *Soil Physics*, third edition. John Wiley and Sons. New York.

Brutsaert, W., G. S. Taylor and J. N. Luthin. 1961. Predicted and experimental water table drawdown during tile drainage. *Hilgardia*, **31**:389–418.

Childs, E. C. 1960. The nonsteady state of the water table in drained land. *J. Geophys. Res.*, **65**:780–782.

Childs, E. C. 1957. The physics of land drainage. *Am. Soc. Agron. monograph* 7 (Ed., J. N. Luthin), 1–78.

Childs, E. C. and N. Collis-George. 1950. The control of soil water. *Advan. Agron.*, **2**:233–272. Academic Press. New York.

Holmes, J. W. 1955. Water sorption and swelling of clay blocks. *J. Soil Sci.*, **6**: 200–208.

Prandtl, L. and O. G. Tietjens. 1934. *Fundamentals of Hydro- and Aeromechanics.* Dover Publ. New York.

Richards, L. A. 1949. Methods of measuring soil moisture tension. *Soil. Sci.*, **68**: 95–112.

Russell, M. B. 1941. Pore size distribution as a measure of soil structure. *Soil Sci. Soc. Am. Proc.*, **6**:108–112.

Russell, M. B. 1942. The utility of the energy concept of soil moisture. *Soil Sci. Soc. Am. Proc.*, **7**:90–94.

Russell, M. B. and L. A. Richards. 1938. The determination of soil moisture energy relations by centrifugation. *Soil Sci. Soc. Am. Proc.*, **3**:65–69.

Taylor, George S. 1960. Drainable porosity evaluation from outflow measurements and its use in drawdown equations. *Soil Sci.*, **90**:338–343.

PROBLEMS

1. A glass tube open at both ends, with an internal diameter of 0.5 mm, is placed vertically in a dish of mercury. What is the depression of mercury in the tube? (Surface tension Hg–air = 470 dynes/cm, the angle of contact = 132°, Hg–glass.)
2. A glass tube open at both ends is inserted into a pan of water. Water rises 2 cm in the glass tube. What is the radius of the glass tube? (Surface tension water–air = 75 dynes/cm, contact angle = 0°.)
3. A soil is placed in a pressure-plate apparatus. The (gauge) air pressure is adjusted to $\frac{1}{3}$ atmosphere. What is the smallest-size (radius) pore that will be drained at this pressure?
4. Repeat this calculation for a pressure of 15 atmospheres.
5. The formula for centrifugal force is given by

$$F = m\frac{v^2}{r}$$

where F is the force in dynes, m is the mass in grams, and v is the speed in cm/sec around a curve of radius r.

(a) Calculate the centrifugal force on 1 gram of soil when moving at 1800 rpm around a circle of radius 50 cm.

(b) How many g's does this represent.

6. A 100-gram sample of dry soil is placed on a porous plate apparatus as shown in Fig. 6-8. The following experimental values are obtained:

T, cm	*Pipette Reading*, cc
0.0	28.0
10.0	27.9
20.0	27.7
30.0	27.3
40.0	26.2
50.0	25.1
60.0	24.0
70.0	23.0
80.0	22.1
90.0	21.6
100.0	21.3

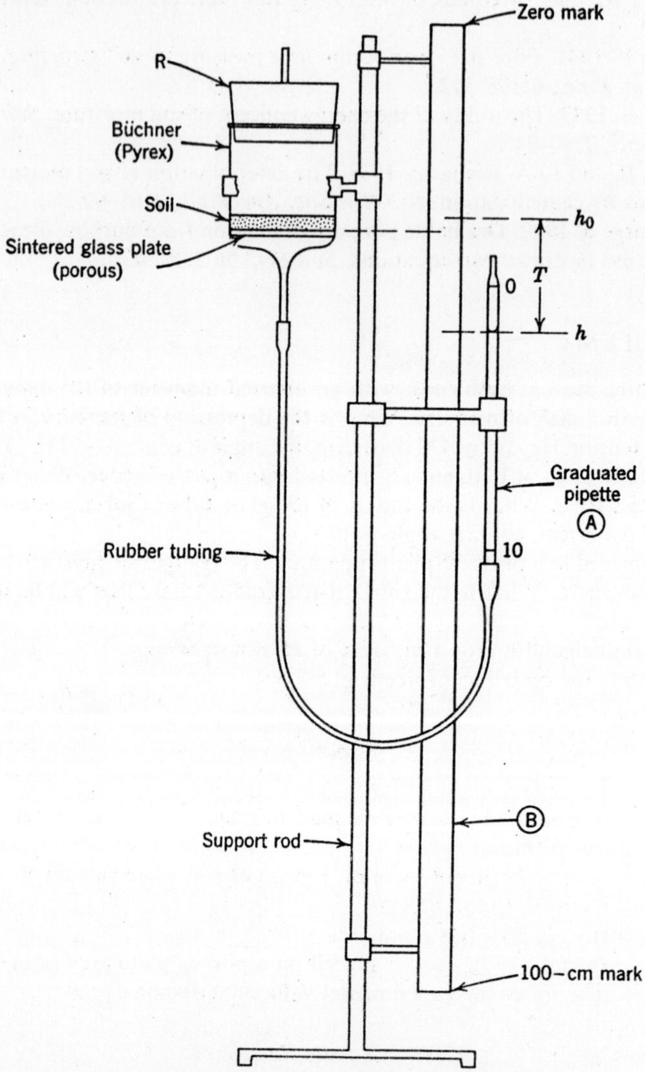

Figure 6-8 Porous-plate apparatus.

At zero capillary pressure the soil contains 35% moisture by dry weight.

(a) Plot the moisture-characteristic curve for this soil.

(b) Assume a bulk density of 1.30 and calculate the moisture concentration for each value of T.

(c) Plot the drainable-porosity curve.

(d) If the water table is initially 15 cm from the soil surface and drops to 45 cm from the soil surface, how much water will drain out of a column of soil 1 sq cm.

Chapter 7 DYNAMICS OF SOIL WATER

FLOW OF WATER THROUGH CAPILLARY TUBES (POISEUILLE'S LAW)

A soil can be considered a collection of capillary tubes. This is only a rough approximation of the actual character of the soil, but much can be learned by making such an assumption. The law of flow of water through capillary tubes has been well known for a long time, and by examining this flow we can deduce certain concepts concerning the flow of fluids through porous media.

Observations by many investigators have shown that the flow of water through pipes or through tubes of capillary dimensions may take place in two different ways. If the velocity of movement is sufficiently low the particles move in parallel elements. The velocity of each element differs only slightly from that of the adjoining one. This type of flow is called laminar, streamline, or viscous flow. When the velocity is sufficiently high the character of the motion changes completely. The orderly laminar arrangement of streamline flow is replaced by a chaos of swirling eddies and vortices, and is known as turbulent flow.

If the flow is laminar, as it is normally considered to be in soils, a certain shearing stress, which is proportional to the velocity gradient and is at right angles to the flow, must be exerted over an area A as shown in Figure 7-1.

In the figure, B and C represent two planes, each of area A in the fluid and parallel to the direction of flow. The planes are X units of distance apart, and the velocities of the flow at the planes are indicated by the relative length of the arrows. The force, F, which causes flow of the upper plane relative to the lower plane is proportional to the area of the planes and to the velocity gradient dv/dx between the planes. It can be expressed mathematically as the formula $F = \eta A \cdot (dv/dx)$ which is known as Newton's law of viscosity. The constant of proportionality η in the equation is called the coefficient of viscosity. It is determined experimentally by measuring the flow produced by a known shearing stress under conditions which render the flow laminar. The poise is a unit of viscosity when the other units are expressed in C.G.S. units. The poise is a rather large unit; one-hundredth of a poise, or one centipoise, is used more frequently.

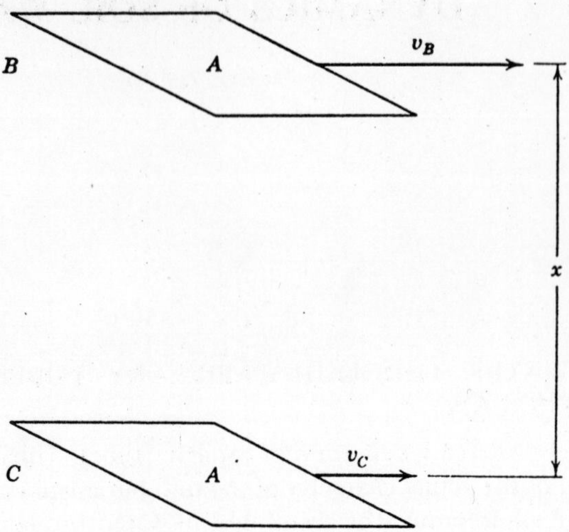

Figure 7-1 Diagram for illustrating Newton's law of viscosity.

We now proceed to make use of the above information in the derivation of Poiseuille's law for flow through a capillary tube. Let us consider the flow through a cylinder of radius R. Let us assume that the fluid is maintained in uniform viscous flow through the length L of the tube by means of a pressure drop ΔP. If we consider a portion of the liquid which is bounded by a circular tube of length L and of internal radius r, the effective force F which is used in overcoming the viscous resistance of the fluid will be $F = \pi r^2 \Delta P$, and it will be acting over the external surface of the tube having an area $A = 2\pi rL$. We now have an expression for the velocity of flow in such a capillary tube:

$$-F = -\pi r^2 \Delta P = \eta(2\pi rL)\frac{dv}{dr} \quad (7\text{-}1)$$

We are interested in the quantity of water flowing through the tube, and if we remember that this quantity is equal to the velocity times the cross-sectional area, πr^2, then the quantity of water flowing is:

$$q = v \cdot A$$

and

$$dq = dv \cdot \pi r^2$$

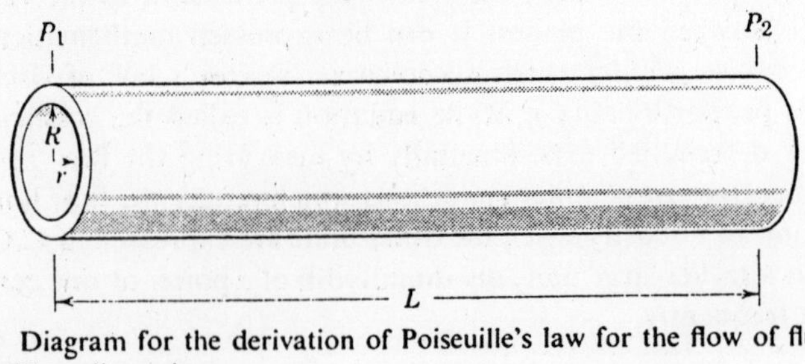

Figure 7-2 Diagram for the derivation of Poiseuille's law for the flow of fluids through tubes of capillary dimensions.

Substituting in equation 1 we have

$$-\pi r^2 \Delta P = \eta(2\pi r L)\frac{dq}{dr}\cdot\frac{1}{\pi r^2} \tag{2}$$

On reduction and separation of variables

$$-\pi r^3 \Delta P\, dr = 2\eta L\, dq$$

When r goes from 0 to R, q goes from Q to 0

$$\int_0^R -\pi r^3 \Delta P\, dr = \int_Q^0 2\eta L\, dq$$

and we have

$$Q = \frac{\pi \Delta P R^4}{8\eta L} \tag{3}$$

which is Poiseuille's law.

Poiseuille's law forms the basis for a number of different equations which have been developed for determining the hydraulic conductivity of the soil from a knowledge of its pore-size distribution. Some of these equations will be considered later in the section dealing with the measurement of the hydraulic conductivity. We can rewrite Poiseuille's law separating out the pressure gradient and the area so that the flow per unit area is given by $Q = (\Delta p/L)\cdot(r^2/8\eta)$. We note that for any particular fluid the flow varies as the square of the radius of the capillary.

FLOW OF WATER THROUGH SOILS (DARCY'S LAW)

Many people have regarded soil as simply a collection of capillary tubes. Poiseuille's law, is then used to make deductions about the system. The capillary tube hypothesis is not entirely valid. Soil is a collection of continuous and sometimes discontinuous pore spaces lying between a solid matrix. The pores are variable in size. Sometimes they are connected with other pores to form a continuous path for flowing water and air. At other times they may be blocked to prevent the movement of liquids. From a microscopic viewpoint the flow of water through soils is immensely complicated. Fortunately an empirical law discovered by Darcy in 1856 describes the macroscopic system. It can be regarded as the fundamental law for flow of water through soils, much as Ohm's law is the basis for the flow of electricity.

A knowledge of Darcy's law is essential to an understanding of the flow of water through soils. It is used not only in drainage problems but also in problems of unsaturated flow. As originally formulated by Henry Darcy the law was based upon an experiment performed in Darcy's laboratory by an English engineer named Ritter.

Darcy's law has been extensively tested and its validity established for the conditions normally encountered in drainage design. Expressed in words the law states that the flow of water through porous material is proportional to the hydraulic gradient and to a factor known as the hydraulic conductivity, k, which is characteristic of porous media. In mathematical symbols Darcy's law is

$$Q = kiA \tag{4}$$

where Q = volume of water per unit time (l^3t^{-1})
i = hydraulic gradient (dimensionless)
A = cross section of flow area (l^2)
k = hydraulic conductivity (lt^{-1})

It is interesting to note the similarity between Darcy's law and other laws which govern physical processes. For example, Ohm's law, which relates to the flow of electricity through a conducting medium, is very similar to Darcy's law in that the flow of electricity is proportional to the voltage gradient and to the specific conductivity of the material. In similar fashion the flow of heat through a conducting solid is also proportional to the temperature gradient and to a property of the material known as the thermal conductivity.

The Energy of Flowing Water

Energy is defined as the ability to do work. In the mechanical system of water and soils we recognize two forms of energy. One is kinetic energy and the other is potential energy.

Kinetic energy is the ability of a mass to do work by virtue of its velocity. The kinetic energy is given by the following equation

$$\text{KE} = \frac{mv^2}{2}$$

The energy per unit weight of fluid is obtained by dividing by the weight of the fluid mg. After division by the weight of the fluid the equation reduces to $v^2/2g$ which is called the velocity head. The velocity of the individual water particles moving through the soil is very small. The kinetic energy due to the velocity can be neglected for the movement of fluids through porous media.

Potential energy is the latent or potential ability to do work. It may be manifested in two different ways. The energy of position, or gravitational energy, is the energy a mass possesses owing to its position above some arbitrary reference plane. With respect to the plane, the mass has a potential energy equal to mgz, where z is the elevation of the mass above the reference plane. This energy is exactly equal to the amount of work required to lift the mass from the datum plane to its present position. The gravitational energy per unit weight of the fluid is obtained by dividing by the fluid weight. The result is called the gravity or positional head and is numerically equal to the distance above a plane of reference.

The second form of potential energy is in the form of pressure, p, which may result from an externally applied force or may be the result of gravity. Any mass containing pressure energy possesses that energy by virtue of contact with other masses containing some form of energy. If we divide the pressure by the specific weight of the fluid we now have the pressure head. The pressure head is the pressure expressed in terms of the height of an equivalent column of water or

$$\text{pressure head} = p/\gamma$$

It is the height reached by a column of fluid under the action of a pressure, p, against gravity.

Since the pressure head and the gravitational head represent potential energy the sum $(p/\gamma + z)$ is called the potential head. In ground-water flow it is known as the hydraulic head or the hydraulic potential. It represents the total energy per unit weight possessed by the water flowing through a porous medium.

Measurement of Hydraulic Head

Hydraulic heads and gradients can be measured in the field. For saturated soils all that is needed is an open-ended pipe placed in the soil to the proper depth. A measurement of the water level in the pipe will give the hydraulic head at the end of the pipe.

Such a pipe is called a piezometer, which means "pressure meter." The pressure head at the end of the pipe is given by the height of water in the pipe. The gravitational head will be the vertical distance from the end of the pipe to a reference plane. The sum of the pressure head and the gravitational head will equal the hydraulic head.

Hydraulic gradients can be measured by putting several pipes side by side, but at different depths below the soil surface, as shown in Figure 7-3. The pressure head at A is h_1, at B is h_2, and at C is h_3. The gravitational heads are respectively z_1, z_2, z_3 and the hydraulic heads are at A, $H_1 = h_1 + z_1$; B, $H_2 = h_2 + z_2$; C, $H_3 = h_3 + z_3$.

The distance from A to B is $z_2 - z_1$, so the hydraulic gradient is

$$\frac{(h_1 + z_1) - (h_2 + z_2)}{z_2 - z_1} = \frac{h_1 - h_2 + z_1 - z_2}{z_2 - z_1} \tag{5}$$

The Velocity Flux and Velocity of Advance

Darcy's law can be rewritten as

$$\frac{Q}{A} = ki = v \tag{6}$$

where Q is the volume of water flowing per unit time through a cross-sectional area, A. Q/A has the dimensions of velocity and is called the velocity flux, v, which is defined as the flow per unit area.

For the flow through a pipe, equation 6 gives the average velocity of the water moving through the pipe. For the flow of water through a porous medium the average velocity differs from the velocity flux, because a substantial part of the flow area, A, is occupied by solid material. The average velocity of the water particles in soils is called the velocity of advance, and

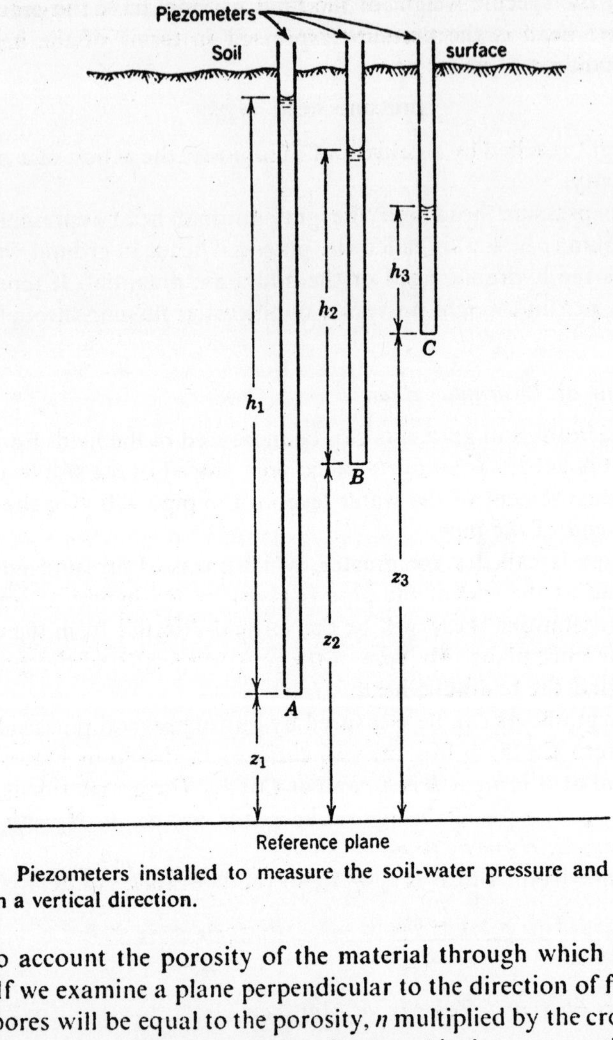

Figure 7-3 Piezometers installed to measure the soil-water pressure and the hydraulic gradients in a vertical direction.

takes into account the porosity of the material through which the water is flowing. If we examine a plane perpendicular to the direction of flow the area of open pores will be equal to the porosity, n multiplied by the cross-sectional area or $A_p = A \cdot n$, where A_p = area of pores, A is the cross-sectional area of the flow section, and n is the porosity.

The average flow velocity will then be given by

$$v_A = \frac{Q}{An} \tag{7}$$

where v_A is called the velocity of advance and Q is the volume of flow per unit time.

Dispersion

The actual velocities of the individual water particles vary from the average given by equation 7 just as velocities vary in the flow of water through a pipe.

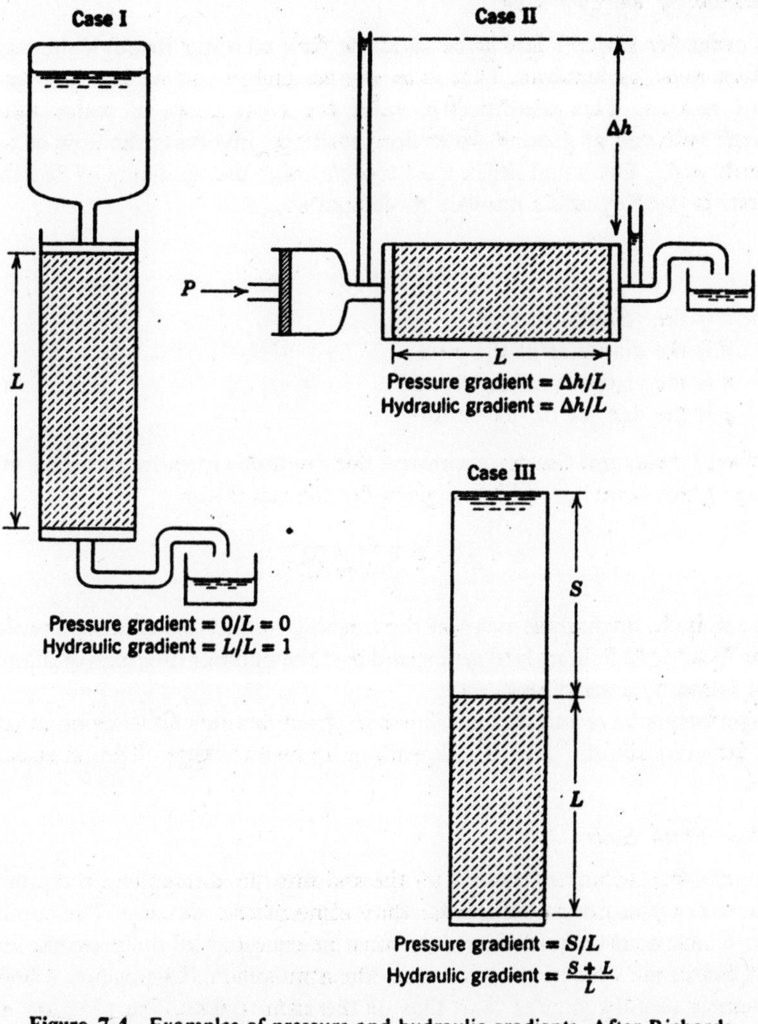

Figure 7-4 Examples of pressure and hydraulic gradients. After Richards.

The central thread of water in each pore of the soil moves faster than the water nearer the walls of the pore.

In addition to the dispersion caused by velocity there is a geometric dispersion caused by the nature of flow through porous media. Because of the repeated branching and subdivision of the pores a drop of dye will spread horizontally as it moves through the soil. This phenomenon was explained by Slichter in 1905. More recently Scheidegger called the phenomenon *dispersion* and presented a mathematical analysis of it. There are many practical implications of dispersion which affect the leaching of soil, the movement of fertilizer through soil, salt water intrusion into aquifers, and so forth.

Limitation of Darcy's Law

In order for Darcy's law to be valid the flow of water through the porous medium must be laminar. That is to say no energy can be lost due to turbulent motion. This condition is valid for most cases of water flowing through soils and of ground-water flow problems involving the flow of water towards wells. The usual index used to determine the tendency of flow to be laminar is the Reynold's number R, defined as

$$R = dv\frac{\rho}{\eta}$$

where v is the velocity of flow
d is the diameter of the pipe
η is the viscosity of the fluid
ρ is the density of the fluid

Fancher, Lewis, and Barnes expressed the Reynolds number in terms of the average grain diameter, which is given by the expression

$$d = \sqrt[3]{\sum n_s d_s^3 / \sum n_s}$$

where d_s is the arithmetic mean of the opening in any two consecutive sieves of the Tyler or U.S. Standard series, and n_s is the number of grains of diameter d_s, as found by a sieve analysis.

Experiments have shown that departure from laminar flow begins at values of R between about 1 and 10, depending upon the range of grain sizes and shapes.

Outflow From Soils

In order for water to flow out of the soil into the atmosphere the pressure in the water phase must be greater than atmospheric pressure. The capillary forces which tend to hold the water must be exceeded by the pressure in the water before the water can escape into the atmosphere. This occurs when the pressure is slightly greater than that of the atmosphere. The pressure must be sufficient to cause drops of water to form at the soil-air interface. The surface-tension forces resisting the formation of drops must be overcome.

Application of Darcy's Law to Some Flow Problems

In the preceding section it was shown that the flow through soil is proportional to the hydraulic gradient and some characteristic of the soil called permeability (hydraulic conductivity).

Soils, as they occur in nature, rarely have uniform permeability (hydraulic conductivity). In most cases the hydraulic conductivity decreases with depth due to the accumulation of clay found in the subsoil of soils. It is of interest therefore to examine the flow through a stratified column of soil to see how Darcy's law may be used to analyze such flow.

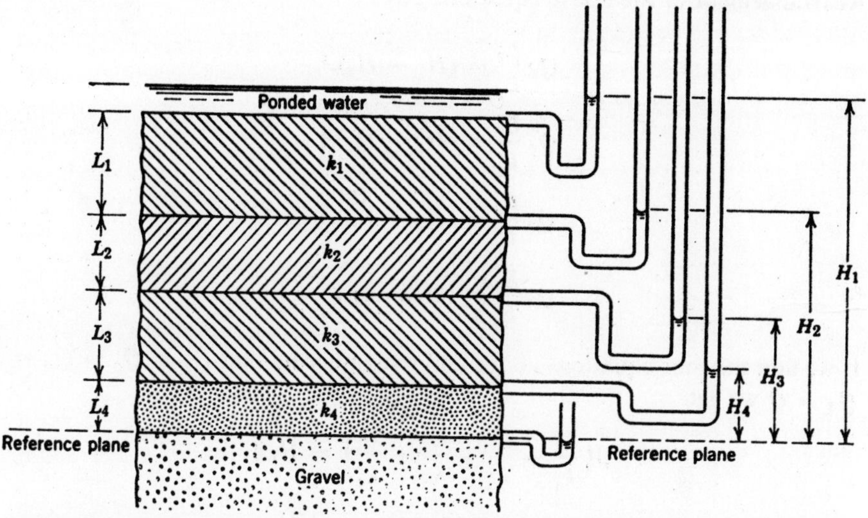

Figure 7-5 Vertical flow through a stratified soil. The water escapes into the gravel at atmospheric pressure. The reference plane is taken to coincide with the upper surface of the gravel stratum.

Let us consider the flow through the column diagramed in Figure 7-5. We take advantage of two relationships in developing an expression for the flow through the column. First we assume that Darcy's law applies to the flow through each section, and second we assume that no water passes through the walls of the container or is stored in the soil. That is to say that the amount of water flowing through any layer is equal to that flowing through any other layer. A similar relationship applies to the flow of water through a pipe of irregular cross section. Another condition which must be imposed is that there is steady state flow which means that the rate of flow is unvarying with time.

With the above mentioned considerations in mind we proceed with the derivation. First we examine the amount of flow through each layer by applying Darcy's law, noting that manometers have been inserted horizontally into the soil at the interface between the various soil layers. Our knowledge of the hydraulic-head gradient stems from the readings obtained with the manometers as indicated in Figure 7-5. The flow through each section is

$$Q_1 = k_1 \frac{H_1 - H_2}{L_1} A$$

$$Q_2 = k_2 \frac{H_2 - H_3}{L_2} A$$

$$Q_3 = k_3 \frac{H_3 - H_4}{L_3} A$$

$$Q_4 = k_4 \frac{H_4 - 0}{L_4} A$$

Rearrangement of the above equations gives

$$Q_1 \frac{L_1}{k_1} = (H_1 - H_2)A$$

$$Q_2 \frac{L_2}{k_2} = (H_2 - H_3)A$$

$$Q_3 \frac{L_3}{k_3} = (H_3 - H_4)A$$

$$Q_4 \frac{L_4}{k_4} = (H_4 - 0)A$$

If we add the four equations above and remembering that $Q_1 = Q_2 = Q_3 = Q_4 = Q$ we get

$$Q\left(\frac{L_1}{k_1} + \frac{L_2}{k_2} + \frac{L_3}{k_3} + \frac{L_4}{k_4}\right) = H_1 A$$

$$Q = \frac{\sum L}{\frac{L_1}{k_1} + \frac{L_2}{k_2} + \frac{L_3}{k_3} + \frac{L_4}{k_4}} \cdot \frac{H_1}{L} A$$

where $\dfrac{\sum L}{\frac{L_1}{k_1} + \frac{L_2}{k_2} + \frac{L_3}{k_3} + \frac{L_4}{k_4}}$ may be called the "apparent" permeability.

Let us suppose that instead of the stratified column of the previous example the water is moving parallel to the direction of the strata. Such situations are very common in the field.

The hydraulic gradient must be the same through the two soil layers. Therefore there will be no flow from one layer to the other. The total flow will be the sum of the individual flows through each layer.

If the flow through 1 is

$$Q_1 = k_1 i A_1$$

and through 2

$$Q_2 = k_2 i A_2$$

then the total flow is

$$Q_1 + Q_2 = k_1 i A_1 + k_2 i A_2$$

or

$$= (k_1 A_1 + k_2 A_2) i$$

The Permeability Factor in Darcy's Law

The word permeability is defined in the dictionary as the quality or state of a porous medium relating to the readiness with which such a medium conducts or transmits fluid. Since this definition is qualitative, we must seek a more precise definition that will withstand rigorous examination from a physical viewpoint. In order to arrive at such a definition it will be appropriate to examine Darcy's law once again. A complete expression of Darcy's law in one

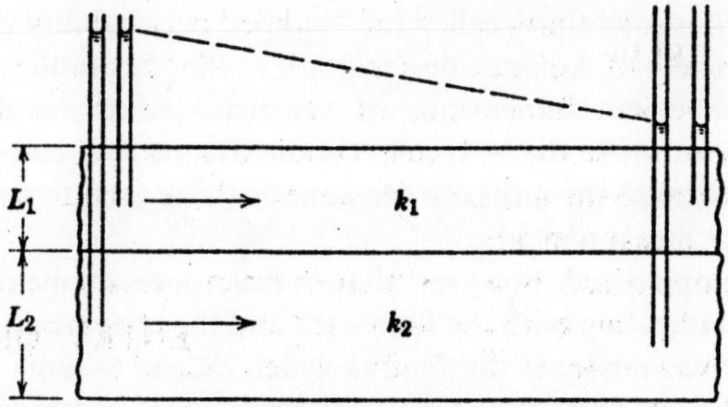

Figure 7-6 Horizontal flow through a stratified soil.

dimension is

$$Q = -\frac{k'g\rho}{\eta} \cdot \frac{dH}{dl} A$$

where ρ is the density of the fluid
g is the acceleration due to gravity
η is the viscosity defined in an earlier section
dH/dl is the hydraulic gradient; here expressed in differential form
A is the cross-sectional area of the flow section
Q is the volume of flow per unit time

Solving for k' in terms of the other quantities in the equation we get

$$k' = Q\frac{\eta}{\rho g} \cdot \frac{1}{dH/dl} \cdot \frac{1}{A}$$

We wish to determine the dimensions of k'

$$k' = (l^3t^{-1})(ml^{-1}t^{-1})(m^{-1}l^3)(l^{-1}t^2)(l^{-2}) = l^2$$

In this case k' is called the *intrinsic permeability* and has the important characteristic of being independent of the fluid or gas used to make the measurements. It might then have the units square centimeters or square feet.

Expressed in words, a medium has an intrinsic permeability of 1 cm^2 when a fluid whose viscosity is 1 poise and whose density is 1 g/cm^3, flows through it with a specific discharge of 1 $(cm^3/sec)/cm^2$ under an impelling force of 1 dyne per gram.

In actual practice we find that an intrinsic permeability of 1 cm^2 is very large compared to measured values, so the unit of length chosen is the micron and the intrinsic permeability is expressed in square microns.

The dimensions of the intrinsic permeability are such that it is not a very convenient unit to use in engineering work. Since we are usually dealing with water as the fluid which is flowing through the porous medium we can simplify the expression for the permeability somewhat by introducing another term, the *hydraulic conductivity* which is defined by the following equation:

$$k = \frac{k'\rho g}{\eta}$$

where k' is the intrinsic permeability.

The symbol k is sometimes called the "lumped permeability constant." The Soil Science Society of America designates it as the "hydraulic conductivity." It has the convenient dimensions of velocity—cm/sec, in./hr, or ft/day. These dimensions make the hydraulic conductivity a very convenient unit of measurement because the units are the same as those used to describe rainfall rates and water measurement.

It must be emphasized, however, that in exact measurements k cannot be given without indicating both the fluid used and the temperature at which the determination was made. If the fluid is water we can assume the density to equal 1, but the viscosity of water varies greatly with temperature. At 10°C the viscosity of water is 1.303 centipoises, at 20°C it is 1.002 centipoises, and at 30°C is 0.789 centipoise. Since the viscosity approaches 1 centipoise at 20°C it is convenient to refer measurements to this temperature by the following relation:

$$k_{20^\circ} = k_{x^\circ} \frac{\eta_{x^\circ}}{\eta_{20^\circ}}$$

The relationship between the hydraulic conductivity and the intrinsic permeability as measured with water may be illustrated by the following example.

Consider a soil for which $k = 10$ cm/hr $= 10/3600$ cm/sec for water at 20°C. Where $\eta = 0.01$ poise, we have

$$k = k' \frac{\rho g}{\eta}$$

and
$$k' = k \frac{\eta}{\rho g} = \frac{0.01 \times 10}{1 \times 980 \times 3600} = 2.8 \times 10^{-8} \text{ cm}^2 = 2.8\ \mu^2$$

Several values for k' in different units are listed below along with corresponding values of k for water at 20°C.

Intrinsic Permeability, k'	*Hydraulic Conductivity, k* (20°C)
2.8×10^{-8} cm²	10 cm/hr
2.8 μ^2	10 cm/hr
1 μ^2	3.5 cm/hr
0.28 μ^2	1 cm/hr
280 mμ^2	1 cm/hr

The above units are by no means all of the units which are used to express the permeability of a porous medium.

In the field of ground water Darcy's law is frequently written

$$Q = PIA$$

where Q is the discharge per unit time
P is the coefficient of permeability
I is the hydraulic gradient
A is the area of cross section

The coefficient of permeability is the rate of flow in gallons per day through a square foot of cross section under a hydraulic gradient of 100% at 60°F.

A unit used in oil production is the darcy. If we express the Darcy equation as

$$Q = \frac{k(P_1 - P_2)}{\eta L} A$$

where η is the viscosity.

If $P_1 - P_2 = 1$ atm
$L = 1$ cm
$A = 1$ cm^2
$\eta = 1$ centipoise
$Q = 1$ cm^3/sec

then k is numerically equal to 1 darcy. Note that in the equation above the gravity-head term is omitted from Darcy's law. In oil fields the pressure head is very large compared to the gravitational head, so it is possible to omit the gravity term with very small error resulting.

CAPILLARY FLOW ABOVE THE WATER TABLE

The flow in the capillary zone above the water table is called capillary flow. It occurs under hydraulic gradients in the same way as flow below the water table, and seems to follow Darcy's law.

The fact that such flow takes place indicates that the water films are continuous throughout the soil and that the pressures are transmitted through these water films.

There is, however, an important difference between capillary flow and flow in the waterlogged zone beneath the water table. The difference arises because of the relationship between the capillary pressure and the hydraulic conductivity. For capillary flow the hydraulic conductivity is called the capillary conductivity. As capillary pressures increase the capillary conductivity decreases. The relationship between the capillary pressure and the capillary conductivity varies from soil to soil.

In general the capillary conductivity falls quite rapidly with increasing capillary pressure. At a capillary pressure corresponding to about one-third to one-tenth of an atmosphere the conductivity is very close to zero. This capillary pressure corresponds roughly to the soil field-moisture capacity.

DISTRIBUTION OF MOISTURE IN A DRAINED SOIL

If sufficient time is allowed and evaporation from the soil surface is prevented the soil moisture will eventually reach an equilibrium state. At equilibrium, to quote Buckingham, "the soil exerts a certain attraction sufficient to hold the water against the action of gravity which tends to drain it perfectly dry."

A plot of the equilibrium moisture content against height in the soil column is called the moisture profile. Childs and Collis-George have pointed

TABLE 7-1 Capillary Conductivity k, mm/day

Capillary Pressure, h (cm)	*Hanwood Loam (Yandera)*	*Banne Sand*	*Camarooka Clay Loam*	*Tuppal Clay*	*Jondergan Clay Loam*
0	750	270	80.0	169.0	60.0
10	350	95	29.5	32.0	18.0
20	180	43.5	11.2	8.0	8.2
40	64	16.3	3.4	2.5	2.6
80	12.5	6.3	0.95	1.03	0.60
150	1.5	2.7	0.23	0.36	0.16
200	0.58	1.55	0.11	0.185	0.082

For the Banne sand the equation relating the conductivity to the capillary pressure is

$$k = \frac{350}{h^{3/2} + 17}$$

For the Camarooka clay loam the equation is

$$k = \frac{560}{h^2 + 80}$$

For k ($h = 0$) the augerhole method was used with 4 to 12 replicates.

This information was obtained from T. Talsma concerning the relationship between the capillary conductivity and the capillary pressure for several Australian soils. All of this data was obtained in the field. The moisture gradients were measured with tensiometers and the water movement was determined from the salt movement. The salt movement was obtained by field sampling to determine the changes in salt with time.

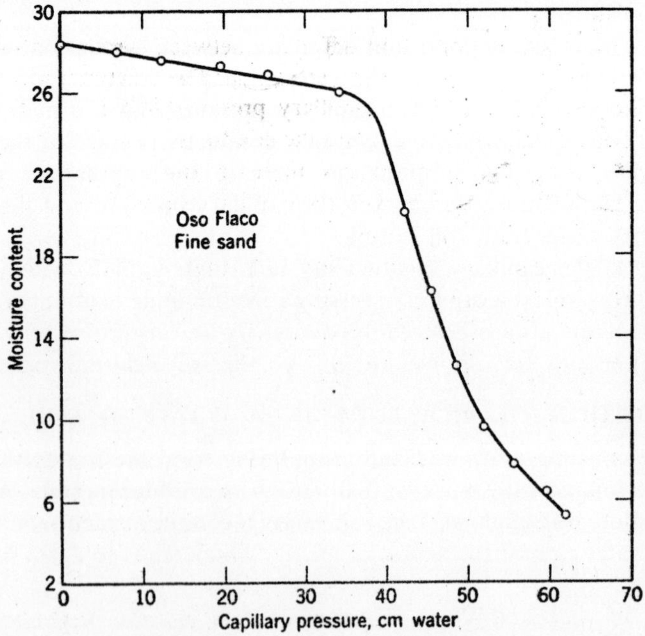

Figure 7-7 Moisture-profile curve.

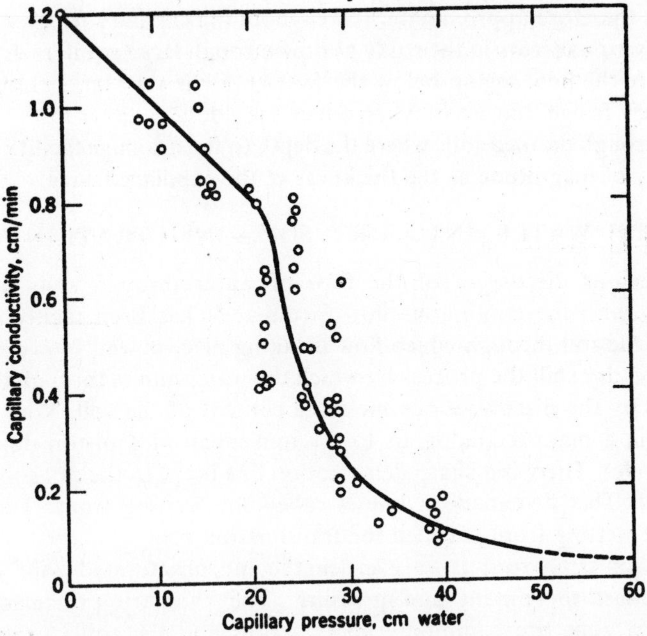

Figure 7-8 Relationship between capillary pressure and capillary conductivity for Oso Flaco Fine Sand.

out the equivalence between the moisture profile and the desorption curve which gives the moisture characteristic.

Since the hydraulic conductivity of a soil is some function of its moisture content, a moisture profile reveals interesting information about the performance of a soil under actual field drainage conditions. In most soils there is a region of uniform moisture content above the water table, with a conductivity that is essentially the same as the hydraulic conductivity of the saturated soil. This region is called the capillary fringe. Childs was the first to use the term in this way, and his usage of the term capillary fringe differs from the common usage of the term for the region wetted by capillary rise from a water table in arid areas.

In sand the capillary fringe may be only 10 to 30 cm thick. For clay loams and clays the thickness may be greater. In other soils having a uniform distribution of pore sizes the capillary fringe may be poorly defined. Whatever the extent of the capillary fringe, it is now evident that it makes a substantial contribution to the flow region in the subsurface drainage of soils.

In some studies of ground-water movement the water table is considered a free water surface. A free water surface is a surface in contact with and in equilibrium with the atmosphere. It is therefore a streamline along which the pressure is atmospheric, and it is a special case of the water table. In the application of the free water surface concept to seepage through soils the effect of movement in the capillary fringe is ignored, and the resultant analysis is therefore approximate. If we ignore the capillary fringe we do not introduce serious errors in the study of flow through large aquifers or through dams where the flow region below the water table is very large compared to the capillary fringe, but we may introduce considerable error in the study of seepage through drained soils where the depth to the drainage facility is on the same order of magnitude as the thickness of the capillary fringe.

ENTRY OF WATER INTO DRY SOIL—INFILTRATION

The previous discussion of the flow of water through soils has dealt exclusively with the continuous flow medium. It has been tacitly assumed that all of the soil through which flow is taking place is wet. However, when rain falls on dry soil, the process by which the water moves through the soil is influenced by the discontinuous moisture content of the soil. Water moves into soils in a manner analogous to the movement of a piston downwards into a cylinder. There is a sharp demarcation line between the dry soil and the wetted soil. This demarcation line is called the wetting front. The region behind the wetting front is called the transmission zone.

The transmission zone is an ever-lengthening unsaturated zone of fairly uniform moisture content and pressure. The capillary pressures in the transmitting zone are commonly about 25 cm and the soil is about 80% saturated.

The infiltration rate, which is determined by ponding water on the soil surface, decreases with time from the commencement of the trial. After a period of time the rate no longer decreases and reaches a constant value.

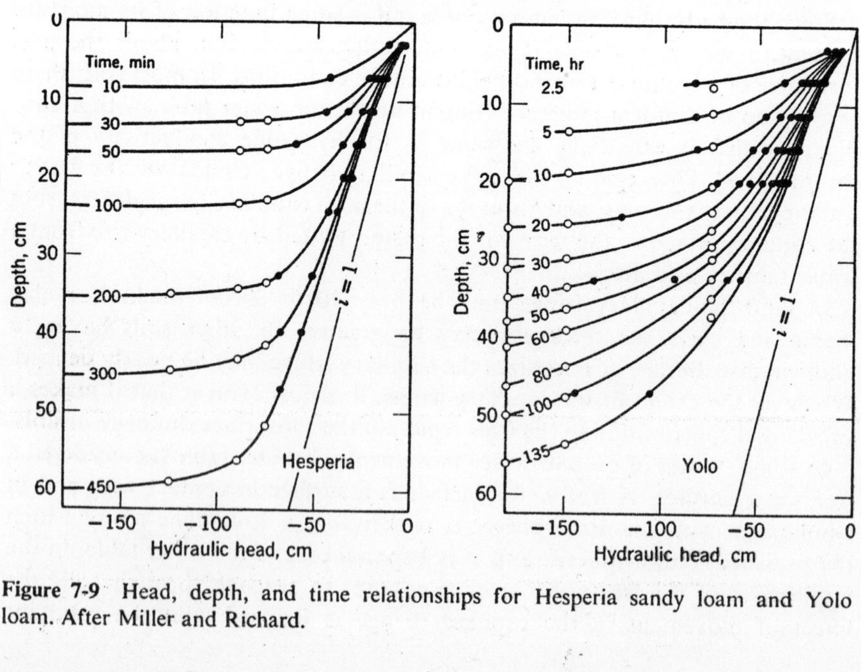

Figure 7-9 Head, depth, and time relationships for Hesperia sandy loam and Yolo loam. After Miller and Richard.

Several empirical equations have been proposed to describe the changing rate of water intake with time. One of these states that

$$i = at^b \tag{8}$$

where i is the rate at which water enters the soil, and t is the time in seconds. The constant a represents the infiltration rate at time = 1, and the exponent b is always negative with values between 0 and −1.

The weakness of the above equation is that the infiltration rate approaches zero at large times. To overcome this difficulty another equation has been proposed to allow for a certain minimum infiltration rate.

$$i = c + at^b \tag{9}$$

where c is the infiltration rate at $t = \infty$.

Equation 8 is better suited to irrigation purposes and equation 9 is used for watershed studies involving rainfall over long periods of time.

REFERENCES

Buckingham, E. 1907. Studies on the movement of soil moisture. *U.S. Dept. Agr. Bur. of Soils Bulletin* 38.

Childs, E. C. 1957. The physics of land drainage. *Agronomy Monograph* **7**:1–78 (J. N. Luthin, editor) American Society of Agronomy, Madison, Wisconsin.

Childs, E. C. 1960. The unsteady state of the water table in drained land. *J. Geophys. Res.*, **65**:780–782.

Childs, E. C., and N. Collis-George. 1950. The control of soil water. *Advan. Agron.*, **2**:233–272. Academic Press, New York.

Christiansen, J. E., A. A. Bishop, F. W. Kiefer, Jr., and Y. Fok. 1959. The evaluation of intake rate constants as related to the advance of water in surface irrigation. *Am. Soc. Agr. Eng.*, Paper 59:713.

Day, P. R. 1956. Dispersion of a moving salt water boundary advancing through saturated sand. *Trans. Am. Geophys. Union*, **37**:595–601.

Fancher, G. H., J. A. Lewis, and K. B. Barnes. 1933. Some physical characteristics of oil sands. Mining Industries Experiment Station, *Penn. State College Bull.* 12.

Luthin, J. N., and P. R. Day. 1955. Lateral flow above a sloping water table. *Soil Sci. Soc. Am. Proc.*, **19**:406–410.

Luthin, J. N., and R. D. Miller. 1953. Pressure distribution in soil columns draining into the atmosphere. *Soil Sci. Soc. Am. Proc.*, **17**:329–333.

Marshall, T. J. 1959. Relations between water and soils. *Tech. Comm.* 50, Commonwealth Bureau of Soils, Harpenden.

Miller, R. D., and F. Richard. 1952. Hydraulic gradients during infiltration in soils. *Soil Sci. Proc.*, **16**:34–38.

Richards, L. A. 1949. Methods of measuring soil moisture tension. *Soil Sci.*, **68**: 95–112.

Richards, L. A. 1952. Report of the subcommittee on permeability and infiltration, committee on terminology. Soil Science Society of America. *Soil. Sci. Soc. Am. Proc.*, **16**:85–88.

Scheidegger, A. 1954. Statistical hydrodynamics in porous media. *J. Appl. Phys.*, 25:994–1001.

Slichter, C. S. 1899. Theoretical investigations of the motion of ground waters. *U.S. Geol. Surv. 19th Annual Report.* Part 2, pp. 295–384.

Slichter, C. S. 1905. Field measurements of the rate of movement of underground waters. *U.S. Dept. Int., Geol. Surv., Water Supply and Irrigation Paper* 140.

PROBLEMS

1. A pressure-head drop of 10 cm of water is maintained across a capillary tube 20 cm long. The tube has a radius of 0.1 mm. What quantity of flow can be expected?
2. Water is ponded on top of a saturated column of soil to a depth of 20 cm. The soil column is 200 cm long. If water drips from the lower end of the column at the rate of 1 cm^3/cm^2/min, what is the hydraulic conductivity? The column has a diameter of 10 cm.
3. A hillside consists of a 3-foot uniform soil over dense granite. The slope of the hillside is 15%. The soil hydraulic conductivity is 4 feet per day.
   (a) What velocity flux can be expected if the soil is saturated?
   (b) If the porosity is 30%, what will be the velocity of advance?
   (c) The soil is saturated for 2 feet above the rock. What will be the seepage per lineal foot into an intercepting ditch at the bottom of the hill?
4. Piezometers are placed side by side in a field at depths of (a) 20, (b) 40, and (c) 60 feet below the ground surface. The pressure heads are 21 feet, 43 feet, and 68 feet respectively.
   (a) What are the hydraulic gradients?
   (b) Which way is the water flowing?
   (c) If the hydraulic conductivity from *a–b* is 2 in./hr what is the conductivity *b–c*?
   (d) What is the "apparent" conductivity *a–c*?
5. An intercepting drain is dug 100 feet from a canal. The water level in the canal is 10 feet higher than the water level in the drain. All the water flows through a horizontal 2-foot soil stratum having a permeability of 1 foot/day. What flow will come out of 100 feet of ditch?
6. The intrinsic permeability of a soil is measured by the air permeameter to be 4 μ^2. What is its hydraulic conductivity at 10°C?
7. Calculate the velocity head for water flowing under unit hydraulic gradient through a soil having a hydraulic conductivity of 0.5 m/day and a porosity of 40%.
8. A layered soil consists of four layers of soil over gravel. Each soil layer is 50 cm thick. The hydraulic conductivities of the layers from top to bottom are k_1 = 0.50 cm/hr; k_2 = 2.30 cm/hr; k_3 = 2.00 cm/hr; k_4 = 4.50 cm/hr. Water is ponded on the soil surface to a depth of 10 cm.
   (a) What is the "apparent permeability" of the soil?
   (b) What is the pressure at the interface between layers (1) and (2)?
   (c) What will be the flow into the ground per unit area?
   (d) Assuming unit hydraulic gradient in a horizontal direction (no ponded water), what will the horizontal flow be per unit area?

9. A soil column consists of two layers of soil. The upper layer is 4 feet thick and has a hydraulic conductivity of 6 inches per hour. The lower layer is 2 feet thick and has a hydraulic conductivity of 1 inch per hour.
   (a) If 2 feet of water are ponded in the soil surface what will be the rate of flow out of the column?
   (b) Will the pressure head increase or decrease with depth in the upper layer? Give a reason.
10. The hydraulic conductivity of a soil at 20°C is 0.024 cm/sec. What is its intrinsic permeability?
11. A 5-foot column of soil is tilted at an angle of 45° to the horizontal. The *average pressure head* at the inflow face is 2.0 feet. The *average pressure head* at the outflow face is 0.10 foot.
   (a) What is the hydraulic gradient?
   (b) What is the pressure gradient?
12. A soil has a hydraulic conductivity of 3 in./hr and a porosity of 35%. For a hydraulic gradient due entirely to gravity what is the velocity of advance?
13. Water is flowing horizontally through three parallel strata having hydraulic conductivities 1.7, 2.4, and 0.50 inches per hour respectively. The strata are each 1 foot thick. If the hydraulic gradient is 2.3 what is the flow per unit width of the soil?

Chapter 8 FUNDAMENTALS OF SEEPAGE ANALYSIS

DERIVATION OF BASIC FLOW EQUATIONS

The linear flow of water through a column of soil is relatively simple and can be analyzed by the application of Darcy's law. However, the two-and three-dimensional flow that occurs in land drainage, seepage under dams, and the flow of water to a plant root is considerably more complicated. Darcy's law is still valid for this flow but it is necessary to derive an equation whose solution will describe adequately the distribution of hydraulic head throughout the region under consideration.

We are now interested in deriving Richard's equation in rectangular coordinates. Laplace's equation will be derived as a special case.

We first examine the flow through a small element of volume. It is convenient to use a small rectangular parallelepiped as the volume element.

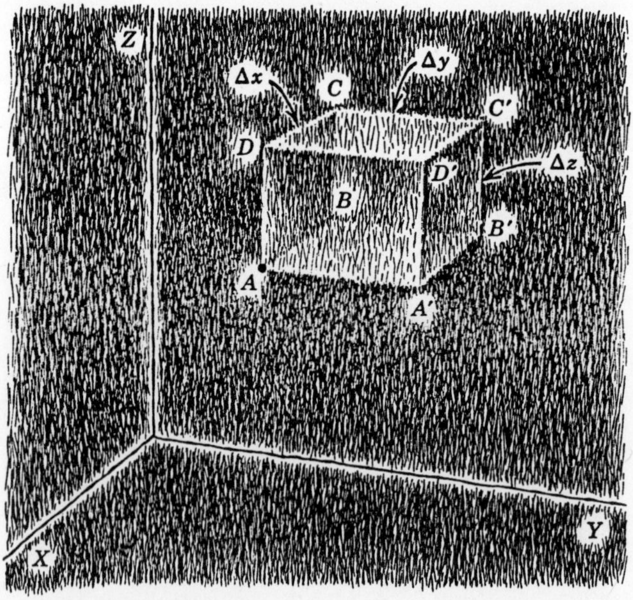

Figure 8-1 Diagram for derivation of flow equation in Cartesian or rectangular coordinates.

In Figure 8-1 a volume element of saturated soil is shown with faces parallel to the xy, yz, and zx planes.

Expressions for the total flow of water into and out of the parallelepiped are needed. If the quantity of water per unit area flowing in the x direction per unit time into the face $BCC'B'$ is q_x, and

$$D'C' = A'B' = AB = DC = \Delta x$$
$$CC' = DD' = BB' = AA' = \Delta y$$
$$BC = B'C' = AD = A'D' = \Delta z$$

then the total water flowing into $BCC'B'$ is $q_x \, \Delta y \, \Delta z$.

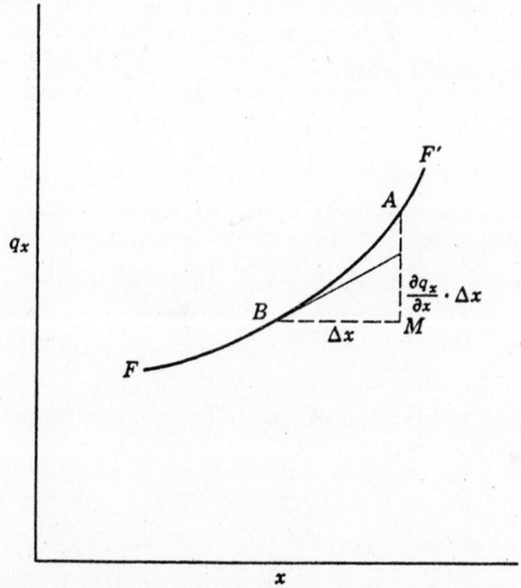

Figure 8-2 The rate of change of q_x in the x direction.

Moving from face $BCC'B'$ to face $ADD'A'$, q_x is changing at a rate of $\partial q_x/\partial x$. At face $ADD'A'$, q_x has changed to $q_x + (\partial q_x/\partial x)\,\Delta x$ approximately.

In Figure 8-2 the geometric significance of the change in q_x to $q_x + (\partial q_x/\partial x)\Delta x$ is shown. The variation of q_x with x is shown by the curve $FBAF'$. The slope of the curve of q_x versus x at B is $\partial q_x/\partial x$. At point A, q_x has increased to $q_x + AM$ where $AM = (\partial q_x/\partial x)\ \Delta x$ approximately and approaches $(\partial q_x/\partial x)\ \Delta x$ as x approaches 0.

Using these results the net (inflow-outflow) flow of water per unit time entering the parallelepiped in the x direction is

$$\text{Inflow–outflow} = q_x \, \Delta y \, \Delta z - \left(q_x + \frac{\partial q_x}{\partial x}\Delta x\right) \Delta y \, \Delta z$$

$$= -\frac{\partial q_x}{\partial x} \Delta x \, \Delta y \, \Delta z \tag{1}$$

If ρ is the density of water, the net mass of water flowing in the x direction is given by

$$-\frac{\partial(\rho q_x)}{\partial x}\Delta x\,\Delta y\,\Delta z \tag{2}$$

The net mass of water flowing in the y direction is

$$-\frac{\partial(\rho q_y)}{\partial y}\Delta x\,\Delta y\,\Delta z \tag{3}$$

and in the z direction

$$-\frac{\partial(\rho q_z)}{\partial z}\Delta x\,\Delta y\,\Delta z \tag{4}$$

If we add equations 2, 3, and 4 we have the net mass of water gained or lost within the parallelepiped

$$-\left[\frac{\partial(\rho q_x)}{\partial x}+\frac{\partial(\rho q_y)}{\partial y}+\frac{\partial(\rho q_z)}{\partial z}\right]\Delta x\,\Delta y\,\Delta z \tag{5}$$

We now introduce a new factor, c, which we will call the moisture concentration and which will be equal to the volume of water per unit volume of soil. The time rate of change of water mass in the parallelepiped is given by

$$\frac{\partial(\rho c)}{\partial t}\Delta x\,\Delta y\,\Delta z.$$

This rate of change of mass of water is equal to equation 5, and the equality can be written

$$-\left[\frac{\partial(\rho q_x)}{\partial x}+\frac{\partial(\rho q_y)}{\partial y}+\frac{\partial(\rho q_z)}{\partial z}\right]=\frac{\partial(\rho c)}{\partial t} \tag{6}$$

Now Darcy's law may be used to simplify equation 6 into a usable form. For general flow that includes capillary flow it is necessary to recognize the dependence of hydraulic conductivity on the soil-moisture pressure. In this case the term hydraulic conductivity is used for flow occurring at pressures both greater and less than atmospheric pressure. It is also assumed that the soil is anisotropic with respect to hydraulic conductivity, so that Darcy's law can be written

$$q_x=-k_x\frac{\partial\phi}{\partial x}\,;\qquad q_y=-k_y\frac{\partial\phi}{\partial y}\,;\qquad q_z=k_z\frac{\partial\phi}{\partial z} \tag{7}$$

A minus sign is used in the differential expressions for Darcy's law, since the hydraulic head drops in the direction of flow, that is, ϕ decreases as x increases when the flow is in the positive x direction.

Substituting the Darcy equations into equation 6 we have

$$\frac{\partial}{\partial x}\left(\rho k_x\frac{\partial\phi}{\partial x}\right)+\frac{\partial}{\partial y}\left(\rho k_y\frac{\partial\phi}{\partial y}\right)+\frac{\partial}{\partial z}\left(\rho k_z\frac{\partial\phi}{\partial z}\right)=\frac{\partial(\rho c)}{\partial t} \tag{8}$$

which is the Richard's equation for the flow of fluids through porous media.

We can obtain Laplace's equation by making the following assumptions. Let us assume that the soil is isotropic with respect to hydraulic conductivity and that the hydraulic conductivity is a constant (i.e., not a function of soil-moisture pressure) then

$$k_x = k_y = k_z = k$$

Also assume that there is no storage or loss of water in the parallelepiped then

$$\frac{\partial(\rho c)}{\partial t} = 0$$

Finally, let us assume that water is incompressible and we have, on simplification of terms,

$$\frac{\partial^2 \phi}{\partial x^2} + \frac{\partial^2 \phi}{\partial y^2} + \frac{\partial^2 \phi}{\partial z^2} = 0 \tag{9}$$

which is Laplace's equation.

The symbol ∇, called "del," is used to denote the differential operator

$$\frac{\partial}{\partial x} + \frac{\partial}{\partial y} + \frac{\partial}{\partial z}$$

and ∇^2, called "del squared," is used for

$$\frac{\partial^2}{\partial x^2} + \frac{\partial^2}{\partial y^2} + \frac{\partial^2}{\partial z^2}$$

We can write Laplace's equation as

$$\nabla^2 \phi = 0 \tag{10}$$

and Richard's equation as

$$\frac{\partial(\rho c)}{\partial t} = \nabla \cdot (\rho k \nabla \phi) \tag{11}$$

COORDINATE SYSTEMS

Rectangular Coordinates

Coordinate systems are frames of reference used to describe geometric objects. It is usually convenient to choose a system of coordinates which bears some simple symmetrical relation to the object or system being described.

Of all the available coordinate systems the rectangular or cartesian coordinates are the most commonly used. These mutually perpendicular axes are used as a frame of reference.

Cylindrical Coordinates

For the analysis of problems having cylindrical symmetry the use of cylindrical coordinates greatly simplifies the solution of a problem, since the angle coordinate drops out of the differential equation.

The analysis of the flow of water towards a well is an important example of the use of cylindrical coordinates.

The symbols used for cylindrical coordinates are shown in Figure 8-3.

The relation between rectangular and cylindrical coordinates is given by the equations

$$x = r\cos\theta, \qquad y = r\sin\theta, \qquad z = z.$$

The element of volume is

$$dv = r\,dr\,d\theta\,dz$$

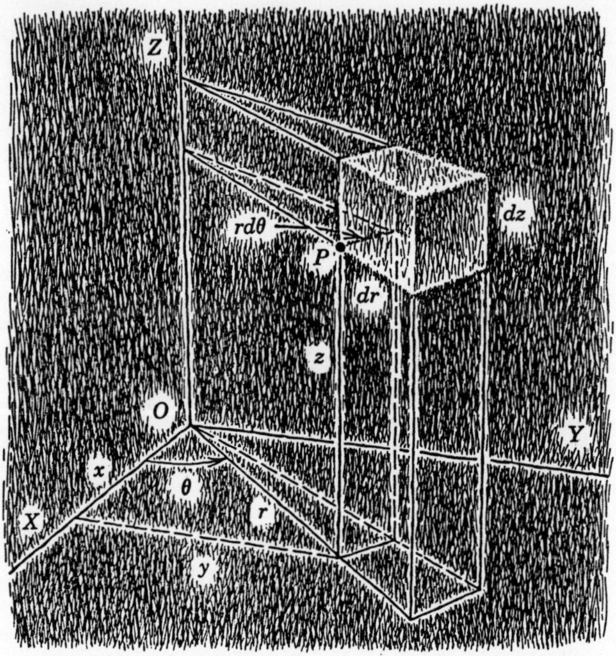

Figure 8-3 Cylindrical coordinates.

Spherical Coordinates

The number of problems involving spherical symmetry is not large. However, it is sometimes possible to get approximate solutions by assuming the existence of spherical symmetry.

The spherical or polar coordinates of a point P are its distance $r = OP$ from the origin, the angle θ between OP and the z-axis, and the angle ϕ in the xy-plane between the x-axis and the plane OPZ. Their relations to rectangular coordinates are expressed by the equations

$$x = r\sin\theta\cos\phi, \qquad y = r\sin\theta\sin\phi, \qquad z = r\cos 0$$

The element of volume is

$$dv = r^2\sin\theta\,dr\,d\theta\,d\phi.$$

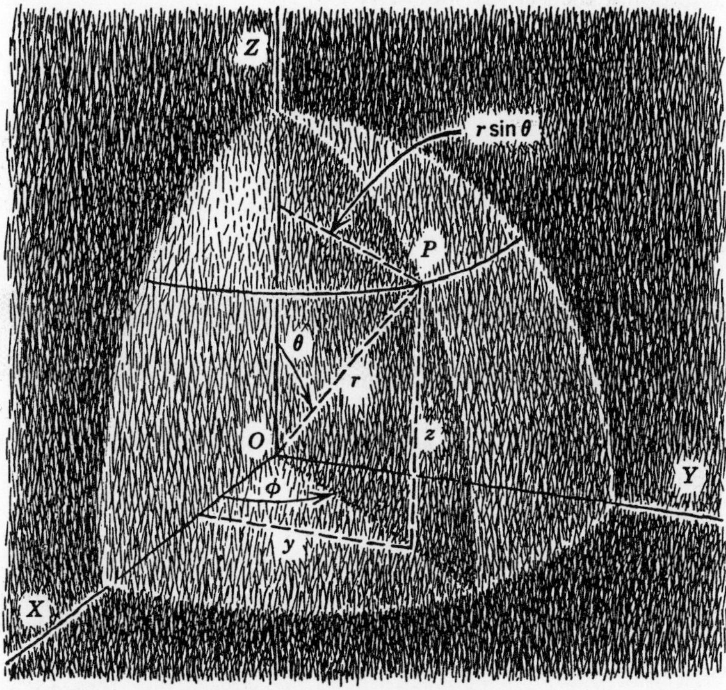

Figure 8-4 Spherical coordinates.

BOUNDARY CONDITIONS

In order to solve a flow problem values of the hydraulic head or values of the hydraulic gradient must be known over the boundaries of the flow region. These values are known as the boundary conditions. Some typical boundary conditions encountered in flow problems are given below.

A Ponded Surface

If water is ponded on a flat soil surface to a known depth, the hydraulic head is the same everywhere on the soil surface. The soil surface is then an equipotential surface and the boundary condition is such that the hydraulic potential ϕ is equal to a constant.

$$\phi = C$$

A Surface of Seepage

Surfaces of seepage occur on the sides of ditches into which water is seeping and on the downstream face of dams through which water is seeping. The pressure head in the water at the seepage surface must be slightly greater than zero for the water to escape into the atmosphere. If the pressure head is zero then the hydraulic head at any point on the seepage surface is equal to the gravitational head at that point, or

$$\phi = z$$

A Line of Symmetry, Streamline, or an Impermeable Layer

The flow on either side of a vertical line drawn halfway between two drains is symmetrical. There is only flow in the direction of the line itself. No flow occurs across the line. The hydraulic gradient at right angles to the line is equal to zero. If n is normal to the line then

$$\frac{\partial \phi}{\partial n} = 0$$

on the line of symmetry. A line of symmetry is also a streamline. Impermeable layers are also lines of symmetry having no flow across them. For streamlines, for impermeable layers, and for lines of symmetry the same boundary condition applies, namely, the hydraulic gradient normal to them is equal to zero.

Free Water Surface

In some phases of ground-water work the water table is regarded as a free water surface. The free water surface concept derives from hydraulics and is defined as a water surface in equilibrium with the atmosphere. It is assumed that the free water surface limits the flow region. There is no flow above this surface. We have already seen that this is untrue for most instances of flow through soils. However, the concept is useful in analyzing the flow through gravels having very small capillary fringes, or wherever the flow region is very large compared to the capillary fringe. The free water surface is considered a streamline having zero pressure. The hydraulic head along the free water surface is taken to be equal to the gravitational head or

$$\phi = z$$

METHODS OF OBTAINING FLOW NETS

A flow net is a plot of equipotentials and of streamlines. Lines of equipotential are defined as having the same potential at every point on the line. Flow occurs at right angles to the equipotential lines in isotropic media. The lines of flow are called streamlines and are orthogonal to the equipotential lines. In anisotropic media the streamlines and the equipotential lines are no longer orthogonal to each other.

Flow nets can be obtained by a mathematical analysis of the problem. This involves a solution of the basic differential equation with the proper boundary conditions. It is beyond the scope of this text to deal with the mathematical procedures that are involved. However, it is significant that there are many problems that are not susceptible to mathematical analysis.

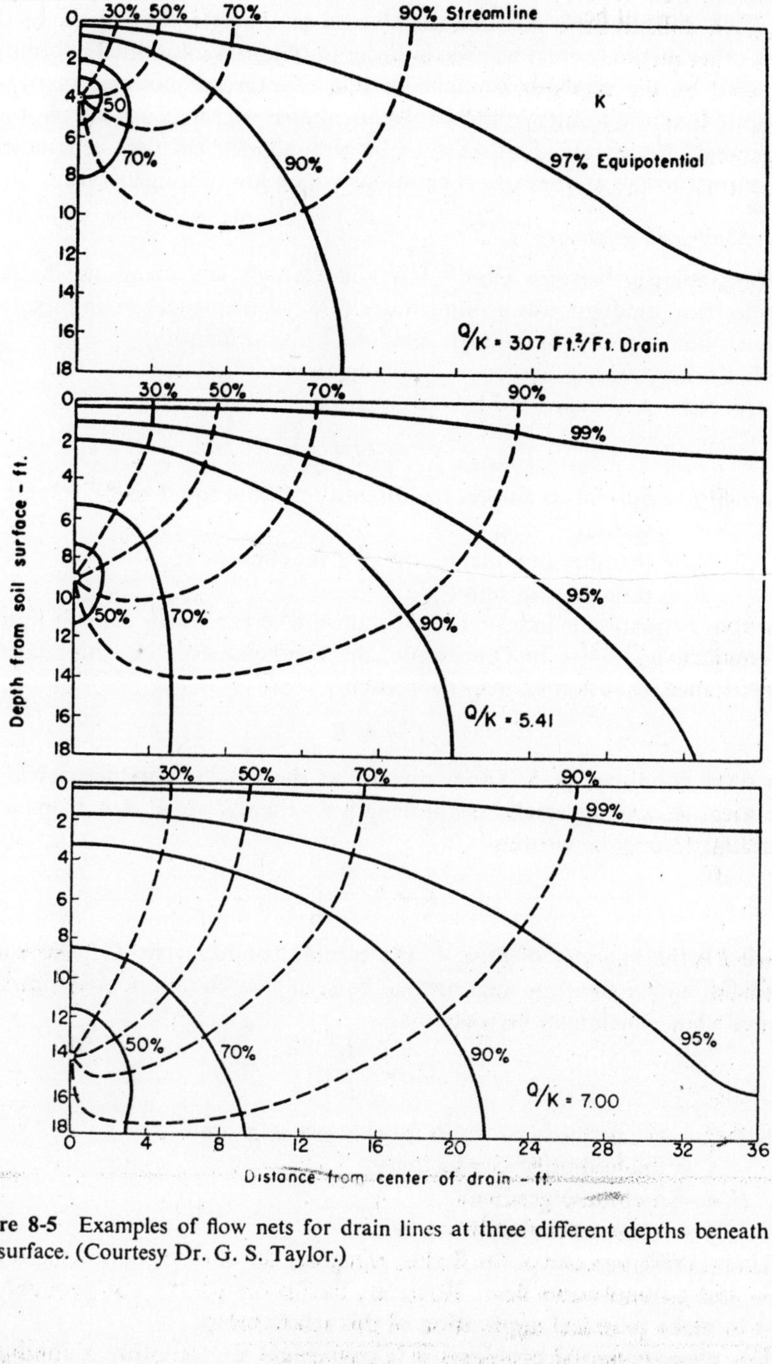

Figure 8-5 Examples of flow nets for drain lines at three different depths beneath the soil surface. (Courtesy Dr. G. S. Taylor.)

Problems that involve irregular boundaries or nonuniform permeability are often difficult or impossible to solve by mathematical analysis. In these cases other methods must be used in order to obtain a solution. The solutions obtained by the methods outlined in this chapter are not necessarily less accurate than the results from mathematical analysis; they are not, however, as general. A separate solution must be obtained for each set of conditions in contrast to the mathematical solutions which are more general.

The Electrical Analogue

The similarity between Ohm's law and Darcy's law forms the basis for the electrical analogue for ground-water flow problems (the term "analogy" means similarity of properties or relations without identity).

Ohm's law, which expresses the fundamental relation for flow of an electric current, is expressed by the equation

$$I = E/R \tag{12}$$

in which I = current in amperes, a quantity of flow equal to 1 coulomb per second

E = pressure in volts, a potential function

R = resistance in ohms

The above equation is in terms of resistance, whereas Darcy's law is in terms of conductivity. Since in Ohm's law, the conductance K' is the reciprocal of resistance, equation 12 may be written

$$I = K'E \tag{13}$$

Since the conductance K' varies directly as the specific conductivity k' and the area, A, and inversely as the length L, then $K' = k' A/L$ from which equation 13 may be written

$$I = k' \frac{E}{L} A \tag{14}$$

Since I is the quantity of flow, k' the specific conductivity, E/L the voltage gradient, and A the flow area, it may be seen that equation 14 is similar to Darcy's law which may be written as

$$Q = k \frac{H}{L} A$$

where Q = the quantity of water flowing per unit time

k = the hydraulic conductivity

H/L = hydraulic gradient

A = cross-sectional flow area.

Slichter (1897) was one of the first to recognize the analogy between electrical flow and ground-water flow. However, Pavlovsky (1922) was probably the first to make practical application of this relationship.

For plane-potential problems it is convenient to use either a conducting salt solution or a sheet of conducting paper. A commercial paper known as

Teledeltos facsimile paper can be used with electrodes painted on with silver paint. The equipotential lines are obtained by applying about 10 volts across the silver electrodes and by using a probe and a high input impedance d-c vacuum-tube voltmeter to trace the equipotential. A silver blueprint-marking pencil can be used to spot the probe positions.

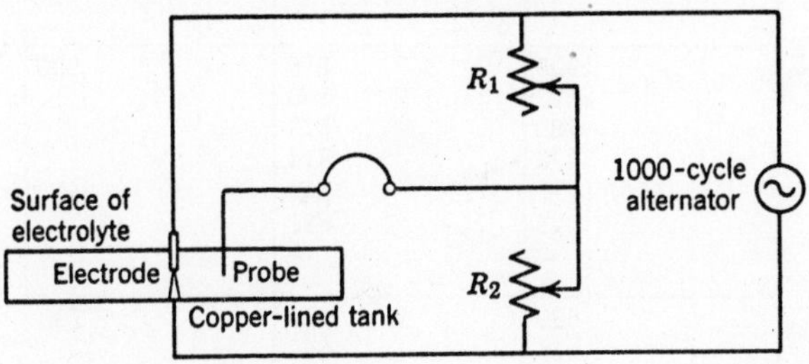

Figure 8-6 A cylindrical copper-lined tank used to study the flow into a well (electrode in center).

Some inherent assumptions for the analogue of the flow of water to a tile line are that water is ponded on the soil surface, the tile line is completely permeable, the tile line is running full with no back pressure, the soil is uniformly permeable, the tile line pictured is one of an array of even-spaced tile lines, and the flow region is underlain by an impermeable layer.

The streamlines can be determined either by drawing them in orthogonally to the equipotentials or by constructing another analogue for the specific purpose of determining the stream function. The streamline analogue is constructed by placing the electrodes on the nonconducting edges in the equipotential analogue, and having a nonconductive area where the electrodes were in the equipotential analogue. A detailed discussion of the streamline case will be found in the section on the resistance network.

Numerical Analysis

ITERATIVE PROCEDURES FOR PLANE POTENTIAL PROBLEMS. Flow problems which involve nonuniform permeability and irregular boundaries are very difficult to solve analytically. Such problems are, however, susceptible to the methods of numerical analysis. Numerical methods form the basis for digital computer solutions as well.

A SIMPLE EXAMPLE—LAPLACE'S EQUATION. Let a square net or grid of uniformly spaced straight lines be drawn over a scale diagram of the region under consideration. In most problems the mesh or side of each square of the net can be chosen so that the boundary passes through the corners of the squares. Once the net has been drawn, known values of the hydraulic head are entered at boundary points and estimated values of the hydraulic head are assigned at each interior point of the net.

In Figure 8-7 values of the hydraulic head at the eight points, A, B, . . . G, H, are taken as known, and estimated values are entered at the top of the columns for interior points a, b, c, and d.

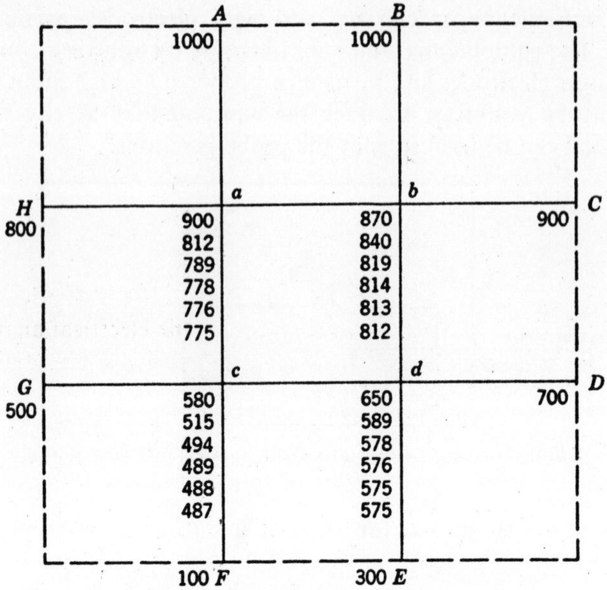

Figure 8-7 Example of numerical analysis.

The interior points of the net are then traversed repeatedly, the value of the head at each point being replaced by the average of its four neighbors. Thus an improved value, ϕ_a', for the hydraulic head at a is

$$\phi_a' = (\phi_A + \phi_b + \phi_c + \phi_H)/4$$

or

$$\phi_a = (1{,}000 + 870 + 580 + 800)/4 = 812$$

Likewise, after ϕ_a' is found, an improved value at b is

$$\phi_b' = (1{,}000 + 900 + 650 + 812)/4 = 840$$

and then

$$\phi_c' = (812 + 650 + 100 + 500)/4 = 515$$

Finally

$$\phi_d' = (840 + 700 + 300 + 515)/4 = 589$$

Now we go back to point a and find a still better value

$$\phi_a'' = (1{,}000 + 840 + 515 + 800)/4 = 789$$

and continue in this way to improve values found for the hydraulic head at interior points. Values obtained after transversing the interior points five times change so little that the problem is regarded as solved, with precision limited by the coarseness of the net used and the number of significant figures used in the calculation. An attempt to improve the solution with this net would be useless, although with a finer net the process might be continued for greater precision.

The basic improvement formula used in the illustration above may be derived as shown below.

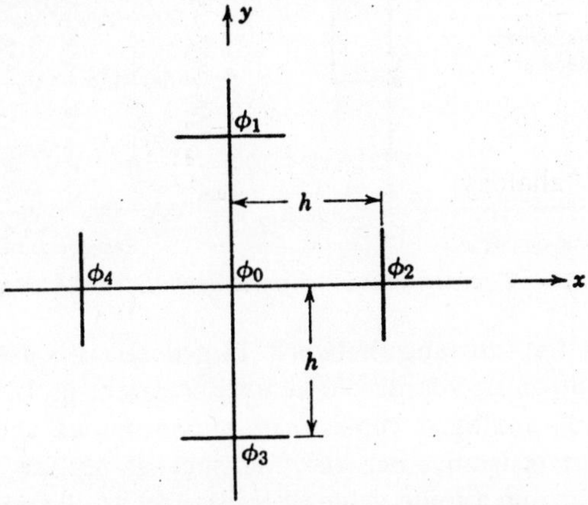

Figure 8-8 Diagram for derivation of iteration formula.

For sufficiently small values of h (see Figure 8-8) the derivative of the function $\phi(x, y)$ with respect to x is approximately

$$\frac{\phi(x + h, y) - \phi(x, y)}{h}$$

and the second derivative is approximately

$$\left[\frac{\phi(x + h, y) - \phi(x, y)}{h} - \frac{\phi(x, y) - \phi(x - h, y)}{h}\right] \Big/ h$$

so that from Figure 8-8

$$\partial^2\phi/\partial x^2 \approx (\phi_2 + \phi_4 - 2\phi_0)h^2$$
$$\partial^2\phi/\partial y^2 \approx (\phi_1 + \phi_3 - 2\phi_0)/h^2$$

Taking the net interval, h equal to unity, Laplace's equation becomes, after reduction

$$\phi_0 = \frac{\phi_1 + \phi_2 + \phi_3 + \phi_4}{4}$$

This is the basic formula for solving Laplace's equation by the iterative method. Special formulas are developed in the same way.

The Resistance Network

The resistance network is probably the most flexible tool devised for solving flow problems except for digital computers. A wide variety of boundary conditions, as well as variations in soil permeability, can easily be

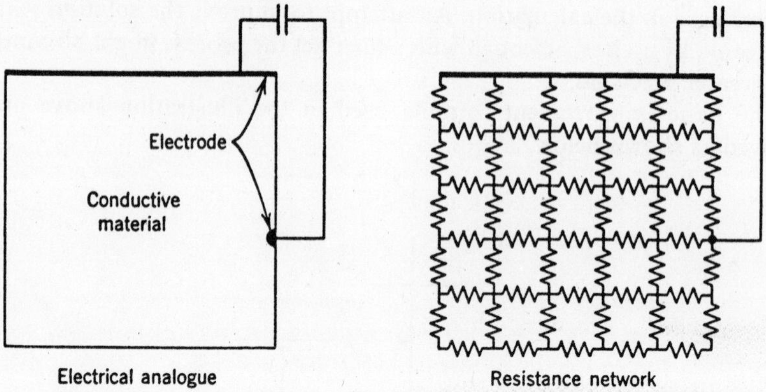

Figure 8-9 Comparison of resistance network to electrical analogue.

simulated with the resistance network. In general, it is used for steady-state problems but it can be adapted to analyze transient problems as well.

The electrical analogue consists of a continuous sheet of conductive material. In the resistance network this sheet is replaced by a network of resistors each having a finite value of resistance, as illustrated by Figure 8-9.

The voltage at any point can be shown to be the average of the voltage at four neighboring points by an application of Kirchhoff's law, which states that the algebraic sum of the currents at a junction equals zero, or symbolically (see Figure 8-10),

$$\sum_{P_1}^{P_4} i_{P_0} = 0$$

Consider a point P_0 in a network of resistors (Figure 8-10). To express the voltage at P_0 in terms of the voltage at the four neighboring points $P_1, P_2, P_3, P_4,$

$$\sum_{P_1}^{P_4} i_{P_0} = 0 = i_1 + i_2 + i_3 + i_4$$

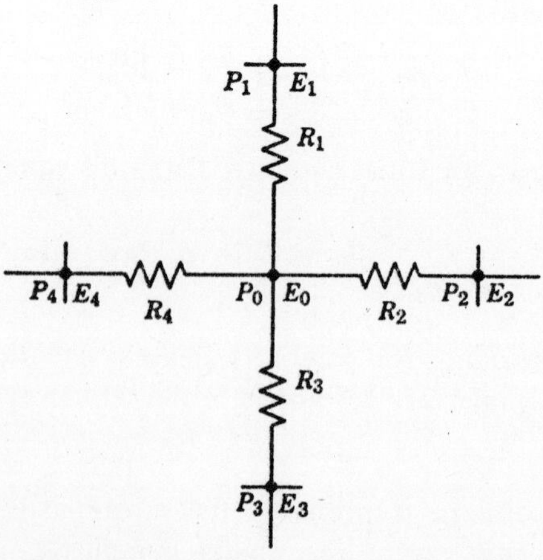

Figure 8-10 Section of resistance network.

When Ohm's law is used to write the current in terms of resistance and voltage, we have

$$\frac{E_1 - E_0}{R_1} + \frac{E_2 - E_0}{R_2} + \frac{E_3 - E_0}{R_3} + \frac{E_4 - E_0}{R_4} = 0$$

If the soil mass is homogeneous, then $R_1 = R_2 = R_3 = R_4$ and we have, after some rearranging,

$$E_0 = \tfrac{1}{4}(E_1 + E_2 + E_3 + E_4) \tag{15}$$

Equation 15 is exactly the same as the basic improvement formula for numerical analysis. The resistance network is thus exactly analogous to the iteration methods, and many of the rules of numerical analysis can be applied to the resistance network after suitable modification.

Boundary conditions can be simulated with the resistance network as follows:

POINTS ON AN IMPERMEABLE BOUNDARY. On an impermeable boundary the first derivative of the hydraulic head taken normal to the impermeable boundary is equal to zero. The average cross-sectional area of the flow section on an impermeable boundary is one-half that of the flow section in the interior. Since the cross-sectional area of the flow section is one-half, the resistance on the boundary must be twice the resistance of interior, since the resistance is inversely proportional to area of flow. Therefore the rheostats on impermeable boundaries are adjusted to twice the resistances of rheostats lying in the interior region.

POINTS ON AN INTERFACE BETWEEN LAYERS OF DIFFERENT PERMEABILITIES. When analyzing the flow between two adjacent points lying on the interface, the assumption can be made that the permeability between the two points is the average of the permeabilities of the layers lying on either side of the interface.

$$R = \frac{2R_1R_2}{R_1 + R_2}$$

where R_1 and R_2 are the basic resistances in the two layers.

STREAMLINES

The resistance network, as well as the electrical analogue, can be made to yield a streamline or flow pattern instead of a plot of equipotentials. The basic arrangement of the network or analogue remains the same because both the stream function and the hydraulic head function are solutions of Laplace's equation. The voltage drop in the network is analogous to both the hydraulic head and the stream function. However, it is necessary to choose a different set of boundary conditions to get the stream function. The change in boundary conditions is based on the fact that the potential function which yields the plot of hydraulic head is an analytic function, and, as a necessary condition for this, it is connected to the stream or flow function by means of

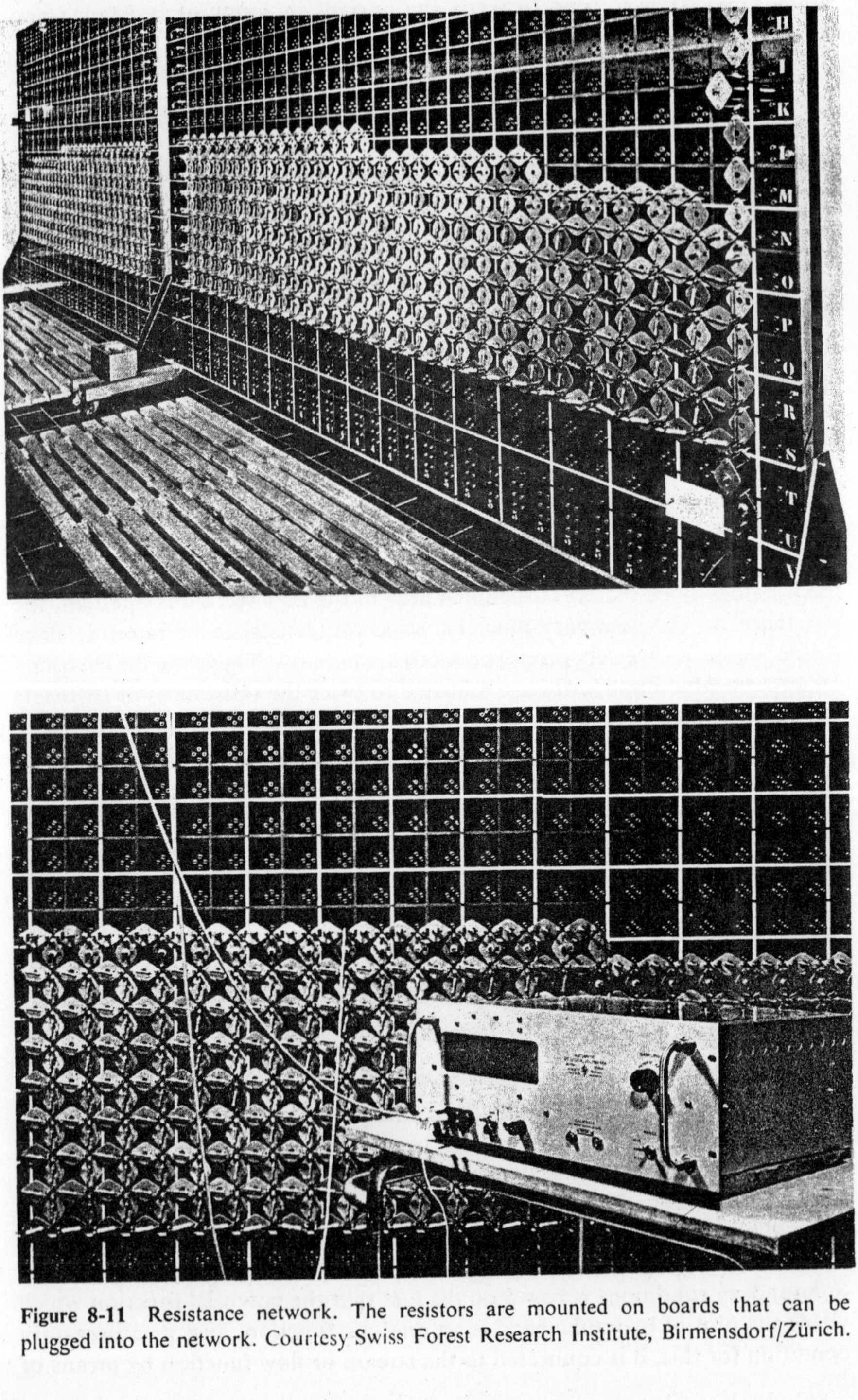

Figure 8-11 Resistance network. The resistors are mounted on boards that can be plugged into the network. Courtesy Swiss Forest Research Institute, Birmensdorf/Zürich.

the Cauchy-Riemann differential equations (cf. E. G. Phillips Functions of a complex variable, p. 12):

$$\frac{\partial \phi}{\partial x} = \frac{\partial \psi}{\partial y}; \qquad \frac{\partial \phi}{\partial y} = -\frac{\partial \psi}{\partial x}$$

where ϕ is the hydraulic head function, ψ is the stream function, and x and y are the usual cartesian coordinates.

Applying the Cauchy-Riemann equations to a vertical impermeable boundary along which $\partial\phi/\partial x = 0$, we get $\partial\psi/\partial y = 0$. This means that along this vertical boundary there is no change in the stream function. The electrical analogue for the stream function will then have an electrode on this boundary, since absence of any change in the stream function indicates presence of an equipotential surface in the analogue. Where the impermeable boundary is horizontal the second Cauchy-Riemann equation can be used in a similar fashion to show the constancy of the stream function along the boundary, and hence the use of an electrode in the electrical analogue for streamline.

If we have an equipotential surface, such as a horizontal water table, $\psi = C$ over the surface, and if x is the cartesian coordinate measured in the horizontal direction, then $\partial\phi/\partial x = 0$. Substitution in the Cauchy-Riemann equations gives $\partial\psi/\partial y = 0$, where y is the vertical Cartesian coordinate. Since ψ does not change at the boundary in the y direction, the boundary can be considered to be impermeable to the ψ function. The equipotential line or surface used for the analogue of the hydraulic head is replaced by an impermeable boundary for the analogue for streamline.

Although the foregoing discussion is concerned with boundaries in either vertical or horizontal directions, the conclusions apply to boundaries in any direction, since Laplace's equation is invariant on rotation of the axes.

The results can be summarized by the statement that an equipotential surface in the hydraulic-head analogue becomes an impermeable surface in the stream analogue. An impermeable surface in the hydraulic head analogue becomes an equipotential surface in the stream analogue.

SURFACES OF SEEPAGE. A surface of seepage is an interface between the saturated soil and the free atmosphere. Since atmospheric pressure is taken as the datum point from which pressure head is calculated, the pressure head over a seepage surface is equal to zero. The total hydraulic head, which is the sum of the pressure head and the gravitational head for saturated soil, is thus equal to the gravitational head over a surface of seepage. The numerical value of the hydraulic head at any point on the surface of seepage is equal to the vertical distance of the point above the reference plane from which the hydraulic head is calculated.

The electrical analogues of seepage surfaces consist of resistance wires or coils of resistance wire over which there is a linear drop in potential. As the resistance of the wire is small compared to that of the conductive solution, most of the current flows through the wire.

A similar arrangement can be used to simulate a surface of seepage on the resistance network. Accurate fixed resistors of low resistance are inserted in the location corresponding to the surface of seepage, and measurements of the desired quantities are made.

ANALYSIS OF TRANSIENT PROBLEMS WITH THE RESISTANCE NETWORK

The successive positions of the water table during transient conditions are found by first finding the potential distribution for an initially known position of the water table. Let us say, for example, that the water table is initially at the soil surface. The next position of the water table is then found by a simple formula derived by Kirkham and Gaskell and based on Darcy's law.

In Figure 8-12 an infinitely small portion of the water table AB is allowed to fall along the streamline AC and BD. Let θ be the slope of the water table and β the angle between the streamlines and the vertical, then it follows that the vertical component of the distance of water-table fall AE is given by

$$AE = AC(\cos\beta - \sin\beta\tan\theta) \tag{16}$$

According to Darcy's law the total distance of fall AC during time T is equal to

$$AC = Tk\frac{\partial\phi/\partial s}{f} \tag{17}$$

where k is the hydraulic conductivity, f is the fraction of the soil which is

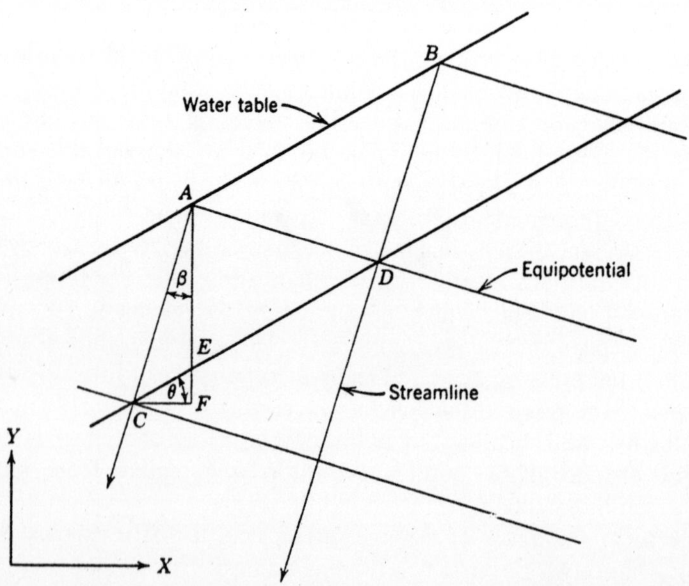

Figure 8-12 Sketch for the derivation of the drawdown formula of Kirkham and Gaskell (1951).

occupied by drainable water, and $\partial\phi/\partial s$ is the partial derivative of the hydraulic potential with respect to path length along AC. Substituting equation 17 in equation 16 results in

$$AE = \frac{Tk}{f}\left(\frac{\partial\phi}{\partial s}\right)(\cos\beta - \sin\beta\tan\theta) \tag{18}$$

Using

$$\frac{\partial\phi}{\partial s}\cos\beta = \frac{\partial\phi}{\partial y} = \phi_y \quad \text{and} \quad \frac{\partial\phi}{\partial s}\sin\beta = \frac{\partial\phi}{\partial x} = \phi_x$$

in equation 18 yields the final equation

$$AE = \frac{Tk}{f}(\phi_y - \phi_x \tan\theta)$$

REFERENCES

Brutsaert, W., G. S. Taylor, and J. N. Luthin. 1961. Predicted and experimental water table drawdown during tile drainage. *Hilgardia*, **31**:389–418.

Kirkham, D., and R. E. Gaskell. 1951. The falling water table in tile and ditch drainage. *Soil Sci. Soc. Am. Proc.*, **15**:37–42.

Liebman, G. 1950. Solution of partial differential equations with a resistance network analogue. *Brit. J. Appl. Phys.*, **1**:92–103.

Luthin, J. N., and R. E. Gaskell. 1950. Numerical solutions for tile drainage of layered soils. *Trans. Am. Geo. Union.*, **31**:595–602.

Luthin, J. N. 1953. An electrical resistance network solving drainage problems. *Soil Sci.*, **75**:259–274.

Richards, L. A. 1931. Capillary conduction of liquids through porous medium. *Physics*, **1**:318–333.

Taylor, G. S., and J. N. Luthin. 1963. The use of electronic computers to solve subsurface drainage problems. *Hilgardia*, **34**:543–558.

Chapter 9 MEASUREMENT OF SOIL PERMEABILITY

RELATIONSHIP BETWEEN PERMEABILITY AND SOIL PROPERTIES

The permeability of soils is a most important physical property since some of the major problems of soil and foundation engineering and agricultural engineering have to do with recognition, evaluation, and solution of drainage problems. These problems include drainage of highways and airports, seepage through earth dams, uplift pressure beneath concrete dams, unwatering of excavated sites to permit construction in the dry, seepage pressures that cause earth slides, failures of retaining walls, and so forth. In all of these the permeability characteristics of soils have a controlling influence on the effective strength properties of the soils and on their responses under stress and hence on stability conditions. Drainable soils will act essentially as open systems with free drainage and fully effective shearing strengths. Soils of low permeability may act as closed systems under rapid application of stress with the development of positive pore pressures and reduction in shearing strength. The determination of the permeability of soils is therefore a most important aspect of soil testing.

In addition the rate at which water and air moves through soils is of considerable importance in agriculture. A knowledge of soil permeability or other soil properties that relate to the rate of fluid movement in both soils and water-bearing aquifers, is needed for the design of drainage facilities. The functioning of drains and drainage wells depends directly upon the capacity or ability of soils and aquifers to transmit water. The flow of fluids through porous media and methods for evaluating soil permeability have occupied the attention of many investigators since the time of Newton, and many significant advances have been made. The basic principles governing the saturated flow of fluids in porous media are now quite well defined and generally understood.

The principal problem from an engineering viewpoint is that of applying the basic theory to evaluate quantitatively the fluid-flow properties of field soils for practical application and design of drainage facilities. The following discussion includes a description of those methods which appear to have the greatest promise for such an evaluation.

DETERMINATION OF PERMEABILITY

Permeability determinations are made on soils and aquifer materials in place in the field and also on samples that are brought into the laboratory. Moreover, measurements may be made either directly by passage of a test fluid through the porous medium, or indirectly by measuring other properties of the medium which are related to permeability.

Direct Measurements

THE DISTURBED SAMPLE. In the use of disturbed soil samples the soil is collected in the field and is packed into permeameter tubes in the laboratory. No special attempt is made to preserve the natural structure of the soil when filling the permeameter. This method has met with success in its application to the structureless soils of the western United States in studies of the effects of various soil treatments on soil permeability. The permeability values obtained are not necessarily related to the permeability of the soil in the field, but serve to indicate the relative effects that various soil treatments might have in the field. Fireman (1944, p. 337) discussed this method in detail:

> In many cases permeability values obtained in the laboratory may not even approximate the percolation rates .... Preliminary tests indicate that, regardless of the correlation between laboratory and field percolation rates, the relative change in permeability obtained in the laboratory as a result of any given treatment is closely correlated with the relative change in percolation rate obtained in the field as a result of a similar treatment.

THE UNDISTURBED SAMPLE. Various investigators have devised methods of obtaining so-called undisturbed soil samples. In general these methods have consisted of inserting a metal cylinder into the soil and removing the cylinder filled with soil. The permeability determination is made directly on the soil-filled cylinder. Large errors enter into the determination because of the presence of root holes and rocks in a sample that is necessarily small. The large variation between replicates makes it difficult to apply the results in the field.

The method is also known as the undisturbed core sample. In certain situations it may be the only available means for getting an estimate of the soil hydraulic conductivity. A typical sampler is shown in Figure 9–1. The sampler is driven or jacked into the ground until the soil is flush with the top of the cylinder. It may be necessary to oil the inside surface of the cylinder to reduce the friction between it and the soil.

The cylinder plus the included soil is then removed from the sampler and is taken to the laboratory for analysis. The permeability may be determined by ponding water on the surface of the soil, or the determination may be made with a falling head permeameter described below.

THE FALLING-HEAD PERMEAMETER. The falling-head permeameter has been in use for many years because of its simplicity. The device consists of a soil

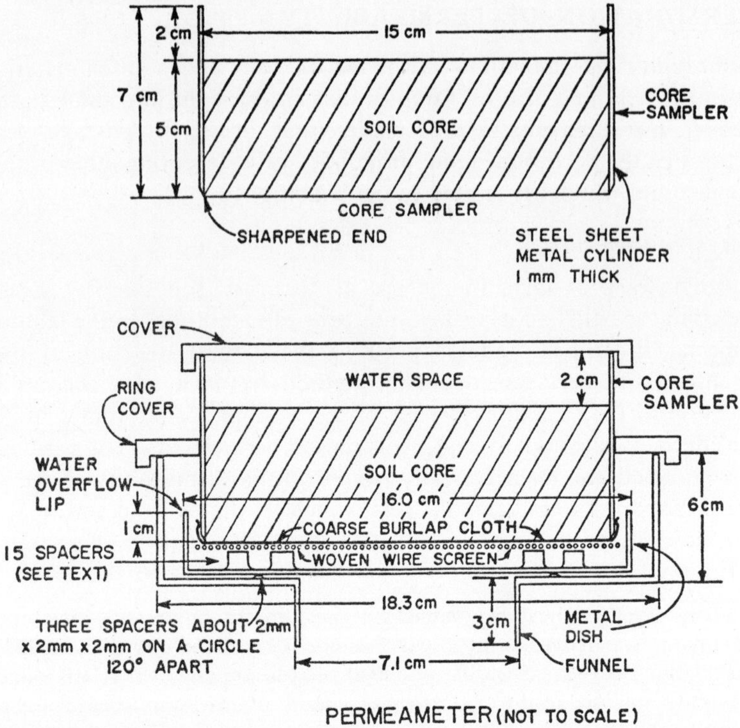

Figure 9-1 Core sampler (*top*) and permeameter assembly (*bottom*), primarily a funnel and core sampler. After Flannery and Kirkham (1964).

column joined to a water column by a U-tube. The water level in the water column is at a higher elevation than the top of the soil and therefore water moves upward through the soil column.

The equation for the falling-head permeameter is developed by considering two expressions for the total flow through the soil.

If Q is the total flow, then dQ/dt is the rate of flow per unit time and, by Darcy's law

$$\frac{dQ}{dt} = \frac{kAH}{L} \tag{1}$$

where k is the hydraulic conductivity

H is the hydraulic head in the water column measured with respect to the upper soil surface

A is the cross-sectional area of the cylinder containing the soil

L is the length of the soil column

The water in the supply column drops from H_0 to H in a time t.

The outflow pressure head on the soil surface remains constant. The total flow Q is also given by

$$Q = aH_0 - aH, \quad \text{or} \quad \frac{dQ}{dH} = -a \tag{2}$$

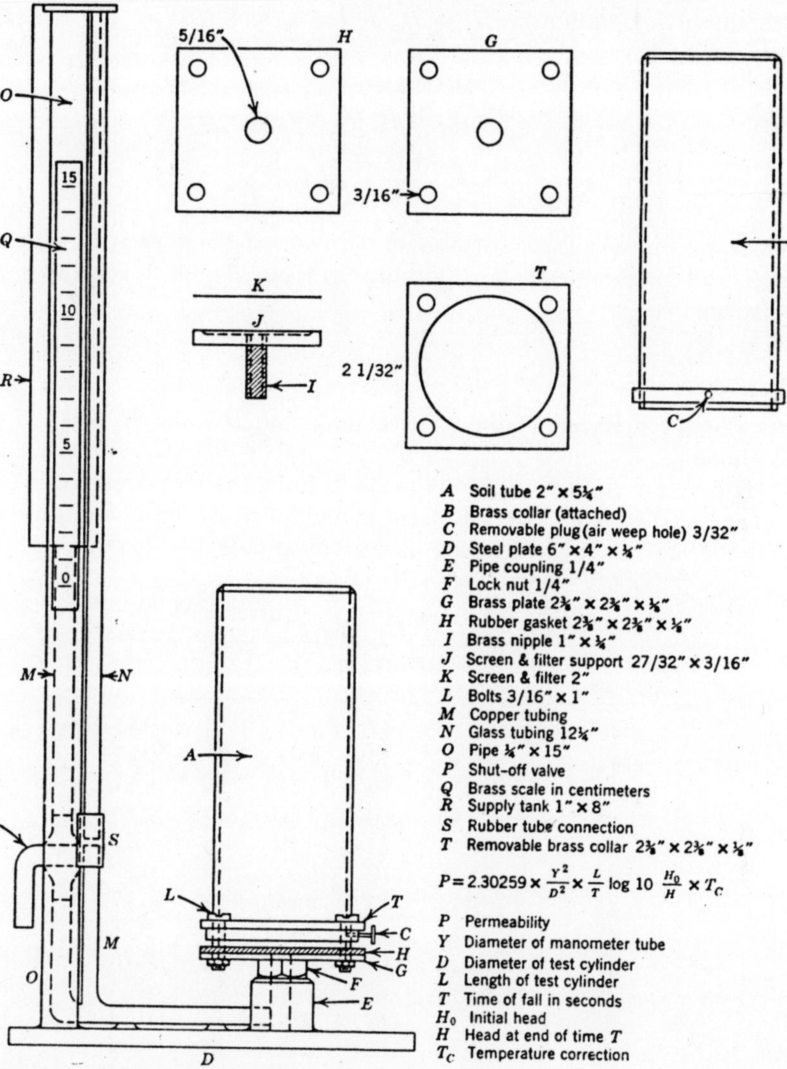

Figure 9-2 Type A. Falling-head permeameter. Soil Conservation Service—Imperial, California.

where a is the cross-sectional area of the supply tube. We now have two expressions for dQ that we can equate. Setting equation 1 equal to equation 2 we have

$$-a\,dH = kA\frac{H}{L}\,dt$$

or, on rearranging,

$$-\frac{dH}{H} = \frac{kA}{aL}\,dt$$

The limits of integration are $H = H_0$ at $t = t_0$ and $H = H$ at $t = t$. On integration we have

$$\ln \frac{H_0}{H} = \frac{kA}{aL}(t - t_0)$$

or

$$k = \frac{aL \ln H_0/H}{A(t - t_0)} \tag{3}$$

which is the formula used to calculate the hydraulic conductivity with the falling head permeameter. The formula can be simplified somewhat by recognizing that

$$\frac{a}{A} = \frac{\frac{1}{4}\pi d^2}{\frac{1}{4}\pi D^2} = \frac{d^2}{D^2}$$

where d is the diameter of the water column and D is the diameter of the soil column.

A temperature correction (Tc) is made to bring the measured value to a standard of 60°F. The observed value is multiplied by the Tc in Table 9-1 to give the value of the hydraulic conductivity at 60°F.

TABLE 9-1 Temperature Corrections

Observed T, °F	*Tc*	*T*	*Tc*	*T*	*Tc*
40°	1.37	57	1.04	74	0.83
41	1.35	58	1.03	75	0.82
42	1.33	59	1.01	76	0.81
43	1.31	60	1.00	77	0.80
44	1.28	61	0.99	78	0.79
45	1.26	62	0.97	79	0.78
46	1.24	63	0.96	80	0.77
47	1.22	64	0.95	81	0.76
48	1.20	65	0.93	82	0.75
49	1.18	66	0.92	83	0.74
50	1.16	67	0.91	84	0.73
51	1.15	68	0.89	85	0.72
52	1.13	69	0.88	86	0.71
53	1.11	70	0.87	87	0.70
54	1.09	71	0.86	88	0.69
55	1.08	72	0.85	89	0.68
56	1.06	73	0.84	90	0.67

Indirect Method

THE KOZENY-CARMAN EQUATION. Water particles moving through the porous medium describe flow paths that conform to individual pore and particle arrangements. In spite of this complexity in the microscopic flow the effective discharge or macroscopic flow can adequately be expressed and described in terms of Darcy's law, which states that the macroscopic or effective flow velocity is proportional to the driving force. Much work has been done in attempting to predict the flow properties of porous media by determining the shape, size, and nature either of the solid particles or the flow channel. The developments along this line in the final analysis depend upon empirically derived constants. The equations that have been derived provide a means of evaluating indirectly the flow properties of porous media. These equations have their greatest application in characterizing the flow properties of single-grain materials such as sand. Of the many equations that have been presented, the Kozeny-Carman equation is the most notable.

The basic form of the Kozeny-Carman equation was originally derived for unconsolidated porous media by Kozeny in 1927 and independently by Fair and Hatch in 1933. It was later reviewed and extended by Carman in 1937 when he wrote that

$$v = \frac{\rho g}{k^1 n S_v^{\,2}} \cdot \frac{n^2}{1 - n^2} \cdot i \tag{4}$$

where v is the velocity flux,

k^1 is the Kozeny-Carman constant, which has an empirically determined value of 5.0 for unconsolidated materials

n is the porosity

S_v is the surface area per unit volume

i is the hydraulic gradient.

This equation has been used widely for determining the surface area of porous materials. However, it may serve as an indirect method for determining permeability for materials for which n and S_v can be evaluated independently.

CHILDS-MARSHALL EQUATION. Pore-size distribution has been correlated with permeability by a number of investigators. Childs and Collis-George 1950 presented a method for calculating the permeability to air and water of the porous medium at all fluid contents, based upon the moisture-characteristic curve which is representative of the pore-size distribution. Their method, however, depends upon a direct permeability determination at complete saturation in order to calculate the permeability at other levels of fluid saturation. The equations are based on the capillary-tube hypothesis and hence Poiseuille's law.

Marshall presented a method of calculating the soil permeability directly from the pore-size distribution, and his equation is

$$k' = \varepsilon^2 n^{-2}[r_1^{\,2} + 3r_2^{\,2} + 5r_3^{\,2} + \cdots + (3n - 1)r_n^{\,2}]/8 \tag{5}$$

where ε is the porosity in cm^3/cm^3 of porous material, k' is the intrinsic permeability in square centimeters and $r_1, r_2 \cdots$ and r_n cm represent the mean radius of pores (in decreasing order of size) in each of n equal fractions of the total pore space.

The measurements are made as follows. The soil-moisture content of the soil is determined as a function of capillary pressure. It is plotted as shown on

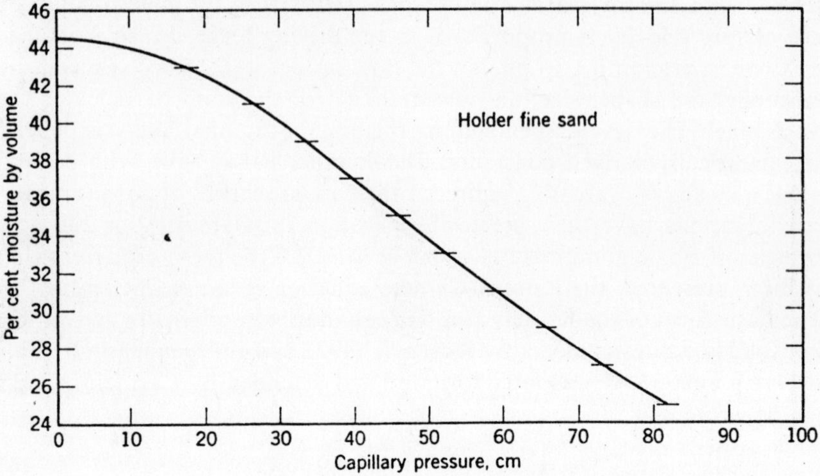

Figure 9-3 Diagram for calculation of soil permeability using the Childs-Marshall method.

TABLE 9-2 Soil Moisture Content for Holder Fine Sand

Porosity Class	*h* (cm)	$1/h^2$	*Multiplier*	*Product*
1	17.4	3.3×10^{-3}	1	3.30×10^{-3}
2	26.2	1.46	3	4.38
3	33.0	0.92	5	4.60
4	39.7	0.64	7	4.48
5	45.8	0.48	9	4.32
6	52.2	0.39	11	4.07
7	58.7	0.291	13	3.79
8	65.3	0.235	15	3.53
9	72.8	0.189	17	3.22
10	81.4	0.152	19	2.89
.				
.				
.				
20				

$n = 20$ Σ prod. $= 38.78 \times 10^{-3}$

$\varepsilon = 0.444$

$k' = (0.444)^2 (20)^2 (38.8 \times 10^{-3})$ cm^2

$k = (k')(2.7 \times 10^2) = 5.15 \times 10^{-3}$ cm/sec at 17.6°C

where 2.7×10^2 converts the intrinsic permeability to hydraulic conductivity. The measured k on the same sample was 3.06×10^{-3} cm/sec.

Figure 9-3 with the moisture content in terms of C, volume of water per volume of soil. The vertical axis on which C is plotted is divided into equal classes. The horizontal lines represent the middle of the class. The capillary pressure associated with each class is then read from the horizontal scale and is recorded as shown in the table below. The capillary pressure is a measure of the square of the pore radius by the expression

$$r^2 = \frac{1}{h^2}\left(\frac{2}{\rho g}\right)^2$$

This expression must be multiplied by $\rho g/\eta$ to convert the intrinsic permeability into the hydraulic conductivity at the temperature of the measurement.

A sample calculation is shown in Table 9-2.

SOIL TEXTURAL RELATIONSHIPS. Soil texture, as estimated by soil survey technicians, has been used as a criterion for drainage design in the Imperial Valley of California. Aronovici (1947) presented a correlation between the silt-plus-clay content of subsoil materials and laboratory values of hydraulic conductivity. This method has the advantage of being simple to apply.

A minimum of measurements is required. The observations are made using the usual soil-survey techniques in evaluating texture. However, this method is no better than the relationships between texture and soil permeability which may or may not be well defined. The ability of the field worker in evaluating the texture of the soil is also involved. It would be expected that such a method would have application in areas such as some of the alluvial valleys of the west, where soil texture rather than soil structure would largely control soil permeability. However, the exchangeable cation status of the soil or the quality of the water may completely vitiate any relationships that might otherwise exist between texture and permeability.

In spite of the limitations listed above a trained soil surveyor can make a rather accurate estimate of the soil permeability from the soil texture and from a visual observation of the soil structure in arid regions.

MEASUREMENT OF HYDRAULIC CONDUCTIVITY IN SITU WHERE A WATER TABLE IS PRESENT

Of all the methods developed for measuring soil hydraulic conductivity in the field, the simplest seems to be to dig an auger hole into the soil below the water table. After first determining the elevation of the water table by allowing the water surface in the hole to reach equilibrium with the soil water, the hole is pumped out to a new elevation of water level in the hole and the rate of rise of the water in the hole is then measured. From these measurements the soil conductivity is calculated. The actual field measurement is simple. It has the advantage of using the soil water for the measurement. The sample used for the measurement is large and the measurement is not greatly affected by the presence of rock or root holes adjacent to the hole. In addition, the measurement largely reflects the horizontal component of the conductivity, which is the important component from a drainage viewpoint.

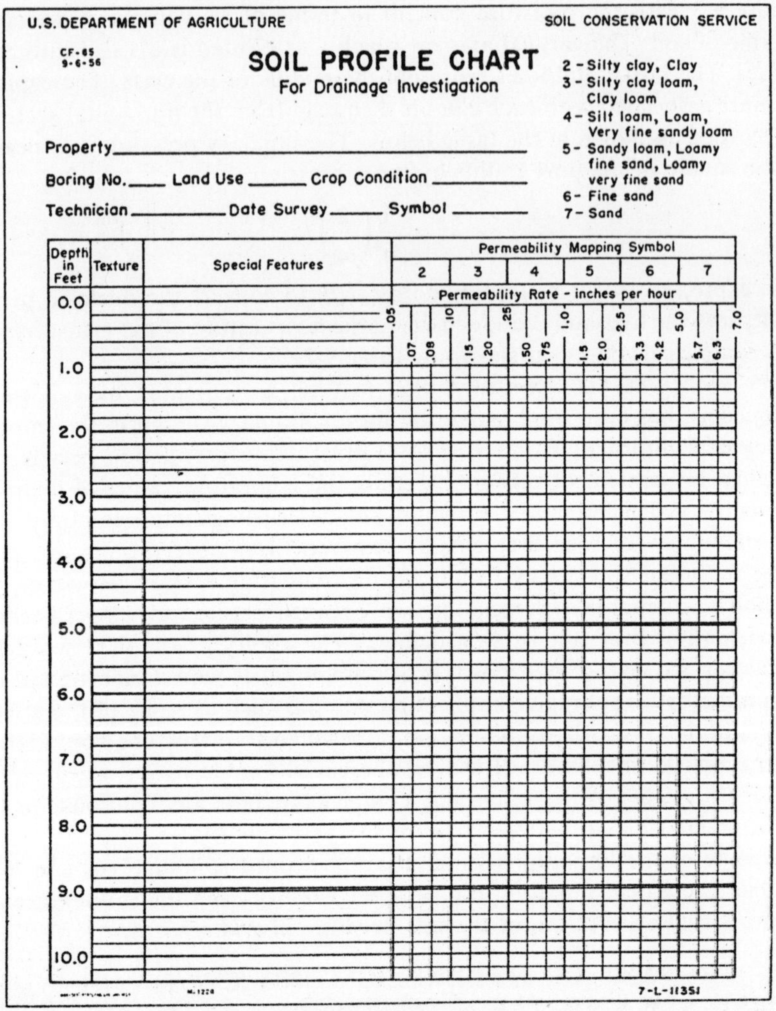

U.S. DEPARTMENT OF AGRICULTURE — SOIL CONSERVATION SERVICE

CF-65
9-6-56

SOIL PROFILE CHART
For Drainage Investigation

Property_____

Boring No.___ Land Use ____ Crop Condition _____

Technician_____ Date Survey____ Symbol _____

2 - Silty clay, Clay
3 - Silty clay loam, Clay loam
4 - Silt loam, Loam, Very fine sandy loam
5 - Sandy loam, Loamy fine sand, Loamy very fine sand
6 - Fine sand
7 - Sand

Depth in Feet	Texture	Special Features	Permeability Mapping Symbol					
			2	3	4	5	6	7
			Permeability Rate - inches per hour					
			.05 .07 .08 .10	.15 .20 .25	.50 .75 1.0	1.5 2.0 2.5	3.3 4.2 5.0	5.7 6.3 7.0
0.0								
1.0								
2.0								
3.0								
4.0								
5.0								
6.0								
7.0								
8.0								
9.0								
10.0								

7-L-11351

Figure 9-4 Soil profile chart for drainage investigation U.S. Department of Agriculture, Soil Conservation Service.

Several different formulas have been developed by various investigators to translate the observed rate of rise of water in the auger hole into the hydraulic conductivity of the soil. Some of these formulas are based upon exact theoretical solutions of Laplace's equation and others are based on approximate solutions. In either case certain physical conditions are assumed and mathematical approximations made to meet the conditions of the test. The utility of the formula will ultimately depend on how accurately the formula gives the hydraulic conductivity in the field. As a result of several thousand measurements in Australia, Maasland and Haskew (1957) conclude that the auger-hole method is accurate. The differences in conductivity obtained between holes was attributed by them to soil inhomogeneity and not to errors in the formulas.

Field methods other than the auger-hole method have also been developed. Kirkham, in 1945, proposed a method based on the flow of water into a cavity beneath the end of a pipe of piezometer. Childs, in 1952, proposed the use of two wells. Water, pumped from one well into the other, is the basis for the determination of soil hydraulic conductivity. The equations used in the various methods are developed in the following sections of this chapter along with a short description of some of the field techniques. It should be noted that the presence or absence of an impermeable layer is important in the choice of a suitable formula. In addition there are special formulas and techniques which can be used for stratified soils.

Single Auger Hole

HOOGHOUDT'S FORMULA FOR HOMOGENEOUS SOIL. Dr. S. B. Hooghoudt made a substantial contribution to drainage design with his development and perfection of the auger hole method of determining the soil hydraulic conductivity.

The method used is very simple and can be done with a small amount of equipment. The procedure is to dig a hole in the soil beneath a water table. After allowing the water level in the hole to come into equilibrium with the water table in the soil, the water is pumped out of the hole and measurements are made of the rate of rise of water in the hole.

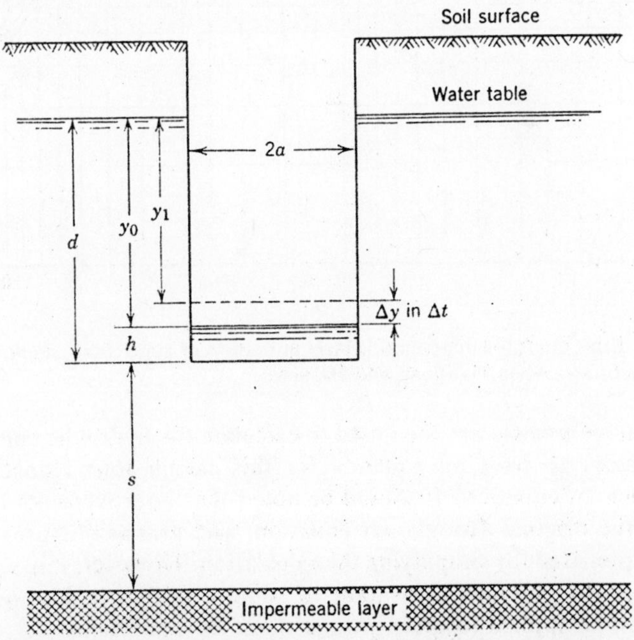

Figure 9-5 Geometrical quantities for auger-hole method—homogeneous soil.

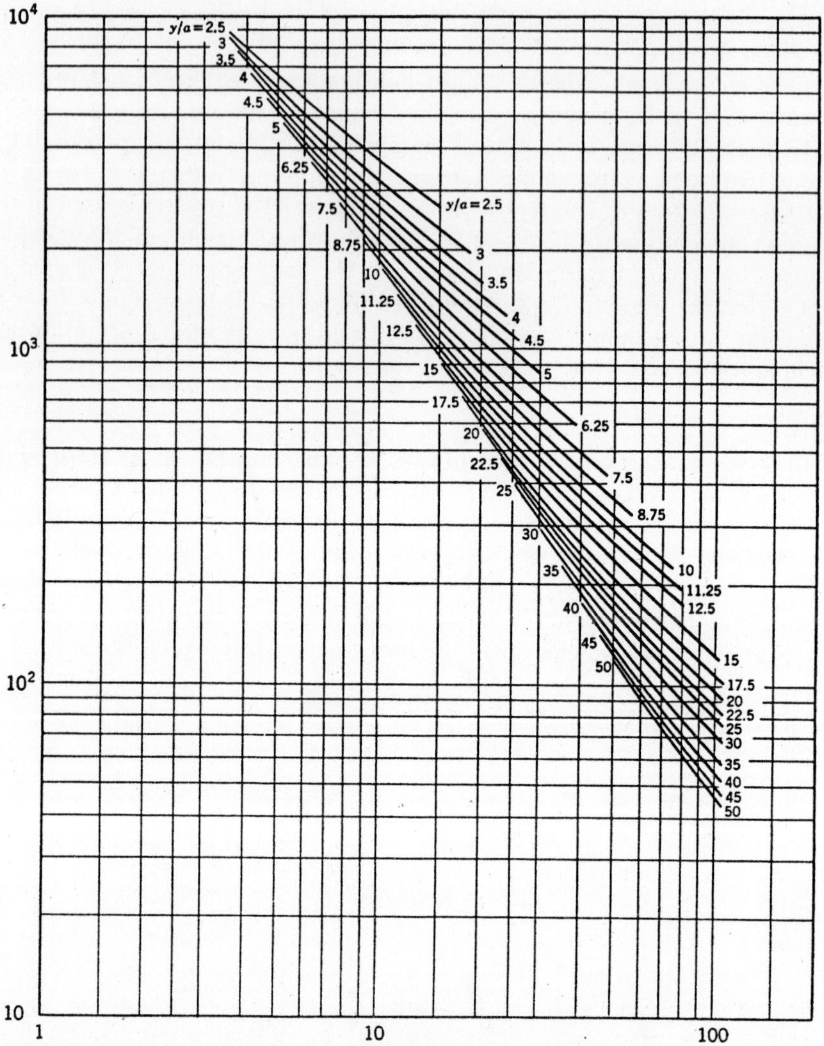

Figure 9-6 Ernst chart for impermeable layer at bottom of auger hole: dy/dt is in ft/sec. k is given in ft/day. After Maasland and Haskew.

These measurements are then used to calculate the hydraulic conductivity. It is necessary to have an equation for this calculation; Hooghoudt has derived such an equation. It should be noted that improvements have been made on the original Hooghoudt equation, and graphs (Figures 9-6, 9-7) have been prepared for simplifying the calculation. However, it is worthwhile considering Hooghoudt's derivation in some detail since his methods of analysis can be applied to other ground-water flow problems.

Hooghoudt developed two formulas, one for use where the auger hole

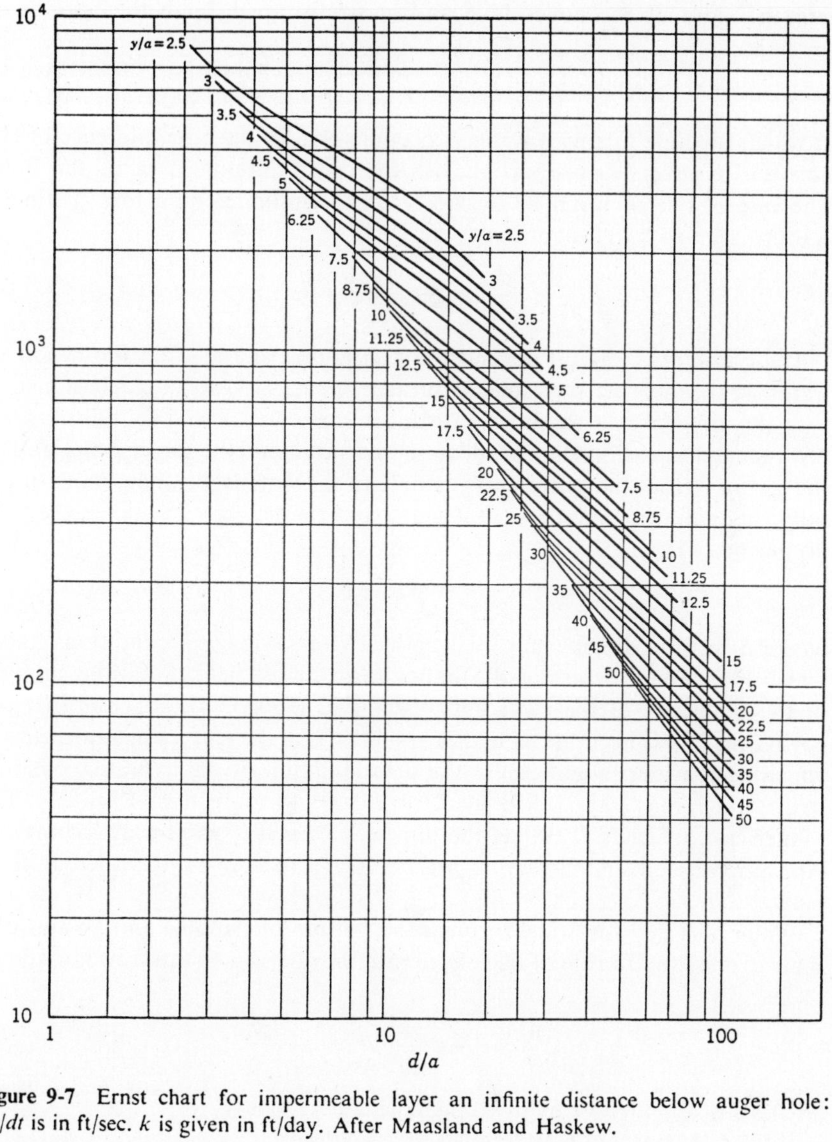

Figure 9-7 Ernst chart for impermeable layer an infinite distance below auger hole: dy/dt is in ft/sec. k is given in ft/day. After Maasland and Haskew.

reaches an impervious layer, and one for use where the impervious layer is at great depths below the end of the auger hole.

One of the assumptions made in his derivation is that the water table is not lowered around the auger hole when water is pumped out of it. The condition is satisfied approximately for a short period of time after the auger hole has been pumped. If however, the auger hole is pumped repeatedly in succession, this condition will not be met.

Another assumption made by Hooghoudt is that water flows horizontally into the sides of the auger hole and vertically up through the bottom of the hole.

Hooghoudt's equation is developed in the following manner. If we assume that the water flows horizontally through the sides of the auger hole, the rate at which the hole fills with water is proportional to the circumference of the hole and is inversely proportional to the cross-sectional area of the hole. The rate of rise of water in the hole due to circumferential flow at time t is thus assumed by Hooghoudt to be

$$\frac{dy}{dt} = -k\frac{2\pi ad}{\pi a^2}\cdot\frac{y}{S} \tag{6}$$

The constant S should depend on a, d, and s, and also on the height of water in the hole at the time of measurement. Hooghoudt determined S with the aid of a controlled experiment in a sand tank and found that S in his assumed equation depended on a, the radius of the auger hole, and d, the distance from the bottom of the hole to the water table, according to the following empirical relation (in which s and the height of water in the hole do not occur):

$$S = \frac{ad}{0.19}$$

where S has the dimension of a length. The experimental conditions under which Hooghoudt determined S only approximated field conditions because of the finite size of the sand tank used, and Hooghoudt suggests that the coefficient 0.19 is accurate to within about 27% of the true value; an accuracy which he considers adequate for the determination of the hydraulic conductivity, which may vary in the field from 0.001 to more than 10 m/day. The numerical coefficient 0.19 has the dimension l and is valid only for meters. Appropriate conversion factors are necessary to convert to other units of length.

Water also flows upward through the bottom of the hole and an approximate expression for rate of rise of water in the hole due to this flow is assumed to be

$$\frac{dy}{dt} = -k\frac{\pi a^2}{\pi a^2}\cdot\frac{y}{S} = -\frac{ky}{S} \tag{7}$$

where S is assumed to be the same constant as before.

If we add equation 6 and equation 7 we derive

$$\frac{dy}{dt} = -\frac{2k\,dy}{aS} - \frac{ky}{S} = -k\frac{2d + a}{aS}y \tag{8}$$

which is the rate of rise of water in the hole due to water entering both the circumference and the bottom of the hole. Integrating equation 8 between the limits $y = y_0$ to $y = y_1$ and $t = 0$ to $t = t$, we have

$$\ln\frac{y_0}{y_1} = k(2d + a)\frac{\Delta t}{aS} \tag{9}$$

or, introducing logarithms to the base 10,

$$k = \frac{2.3aS}{(2d + a)\,\Delta t} \log_{10} \frac{y_0}{y_1} \tag{10}$$

When the auger hole terminates on an impermeable layer the vertical flow of water through the bottom equals zero and equation 10 becomes

$$k = \frac{2.3aS}{2d\,\Delta t} \log_{10} \frac{y_0}{y_1} \tag{11}$$

Since $S = ad/0.19$ we can introduce it in the above equations. Then, for an auger hole on an impervious layer we have

$$\begin{aligned} k &= \frac{2.3(a)^2 86{,}400}{(2) \cdot (0.19)} \frac{\log_{10} y_0/y_1}{\Delta t} \\ &= 523{,}000a^2 \log_{10} \frac{y_0/y_1}{\Delta t} \end{aligned} \tag{12}$$

where the radius of the hole, a, is measured in meters, and Δt in seconds; k is now given in meters per day. The factor 86,400 converts seconds into days.

An example will serve to clarify the use of the equation expressed in metric units.

In Figure 9-5

$$\begin{aligned} y_0 &= 85 \text{ cm} \\ y_1 &= 82 \text{ cm} \\ a &= 3 \text{ cm} \\ \Delta t &= 20 \text{ sec} \end{aligned}$$

then, substituting in equation 12, we have

$$k = 523{,}000 \left(\frac{3}{100}\right)^2 \frac{\log_{10}(85/82)}{20}$$

$$k = 0.358 \text{ meters per day}$$

The conversion factor to change meters per day into inches per hour is 1.64.

ERNST'S CHARTS. Ernst made a detailed study of the flow into an auger hole and developed graphs which are easier to use than Hooghoudt's formula. If we insert the values for the above problem into the graph, we have, using the graph in English units,

$$\begin{aligned} y/a &= 83.5/3 = 27.8 \\ d/a &= 100/3 = 33.3 \end{aligned}$$

The amount of rise is 3 cm, which is equal to 0.0985 feet.

We first draw a line between the top and bottom scales for $d/a = 33.3$ to intersect the $y/a = 25$ and 30. After interpolation a horizontal line to the c-axis gives us a value of about $C = 260$. Then

$$k = C\frac{\Delta y}{\Delta t} = 260\,\frac{0.0985 \times 12}{20 \times 24} = 0.64 \text{ in./hr}$$

Figure 9-8 Measuring the soil permeability by the auger-hole method.

which compares with $0.358 \times 1.64 = 0.596$ in./hr determined by using Hooghoudt's formula.

PERMEABILITY OF LAYERED SOILS BY AUGER HOLE METHOD. *Ernst's method.* When the profile consists of two layers having an appreciable difference in conductivity, Ernst proposes that the hydraulic conductivity of each layer be determined by digging two auger holes of different depths. The bottom of the first hole should be approximately 10 cm above the lower layer. The second hole should extend well into the lower layer. If there is a third layer, the bottom of the second hole should stay above that layer. Since the formulas are based on $d - h$ greater than 15 cm, $d - h$ should be greater than 15 cm. Ernst derives the following equation

$$k_1 d_1 + k_2(d_2 - d_1) = k d_2 \tag{13}$$

Where k_1 is the conductivity as measured in the first hole, k_2 is the conductivity of the lower layer, and k is a mean value of the conductivity for the two layers as measured by the rate of rise in the second hole. The conductivity

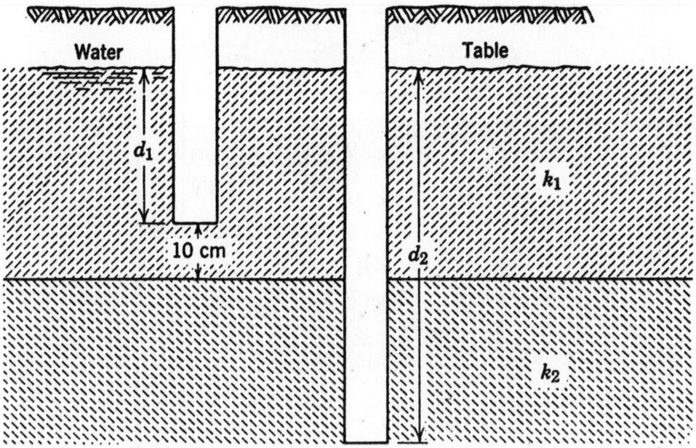

Figure 9-9 Ernst method for layered soil.

of the lower layer, k_2, is computed from the above equation. This equation gives fairly reliable results if k_2 is greater than k_1. If k_2 is very much less than k_1 the equation may give negative results for k_2. If accurate values are desired for k_2 the measurements in the second hole should be delayed until the water table has lowered to a position below the interface of the two layers.

There are two assumptions made in deriving the equations for layered soils. The first one is that the stream lines are horizontal and are independent of the way in which the permeability of the profile changes.

The second assumption is that the amount of water flowing out of each layer depends only on the permeability of the layer out of which it flows. Both of these assumptions are questionable. Ernst, however, indicates that the resulting error is not greater than about $\pm 10\%$.

The Two-Auger-Hole Method—Childs

Childs has proposed a method for nonlayered soils employing two auger holes rather than one. The two holes are of equal diameter and penetrate to the same depth below the water table, preferably to an impermeable layer if one exists. Water is pumped at a steady rate out of one well and carried by a hose into the other creating a small hydraulic head difference between the levels of water in the holes. If Q is the pumping rate. ΔH the hydraulic head difference between the two holes, L the length of each well beneath the water table, a the radius of each hole, and b the distance between their vertical axes, the hydraulic conductivity k is given by

$$k = \frac{Q}{\pi L \, \Delta H} \cosh^{-1} \frac{b}{2a} \tag{14}$$

This equation is valid only if the holes penetrate to an impermeable layer. In the event that the auger hole does not reach an impermeable layer, an "end correction" must be applied to compensate for the flow entering the

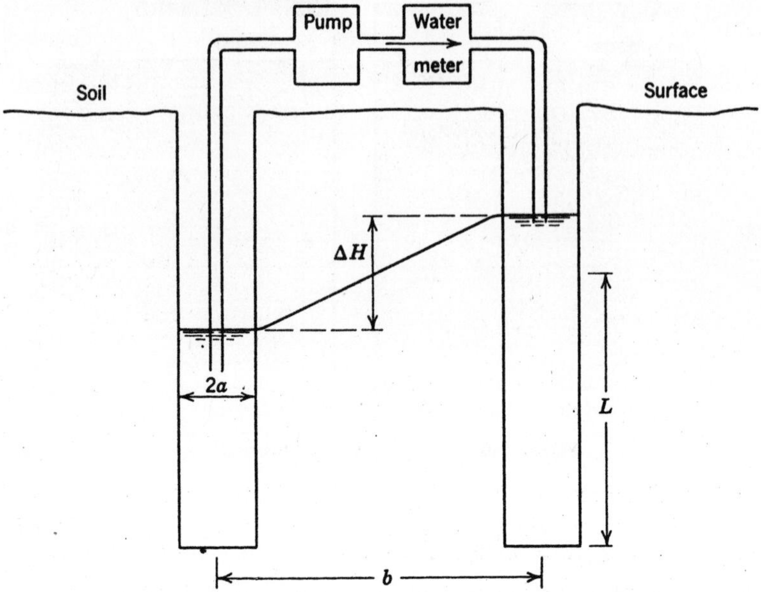

Figure 9-10 The two-auger-hole method of Childs.

end of the auger hole. The end effect may be regarded as a flow which extends the length of the auger hole and depends on the depth to the impermeable layer as well as on the dimensions of the hole. An addition of some 20 cm to the measured depth is suggested by Childs as an appropriate end correction for holes of the radius he used. In addition the effective flow region between the two wells is enlarged by the flow which occurs in the capillary fringe. Once again it is possible to compensate for this flow by extending the effective length of the auger hole. Adding 5 cm to L will usually be adequate, although it is possible to make an estimate of the capillary fringe in the field and to take half of the thickness of the capillary fringe as the fringe correction.

The Pipe-Cavity Method

Kirkham proposed that tubes or pipes be driven into the soil below a water table either with or without a cavity at the end of the tube. The soil would be augered out of the tube, the water table allowed to establish itself, and then water pumped out of the tube to measure the soil conductivity. The rate of rise of the water in the tube would then be transformed into the soil conductivity by the use of suitable equations which were developed by Kirkham.

The equation developed by Kirkham states that

$$k = \frac{\pi r^2 \ln (y_0/y_1)}{S(t_2 - t_1)} \tag{15}$$

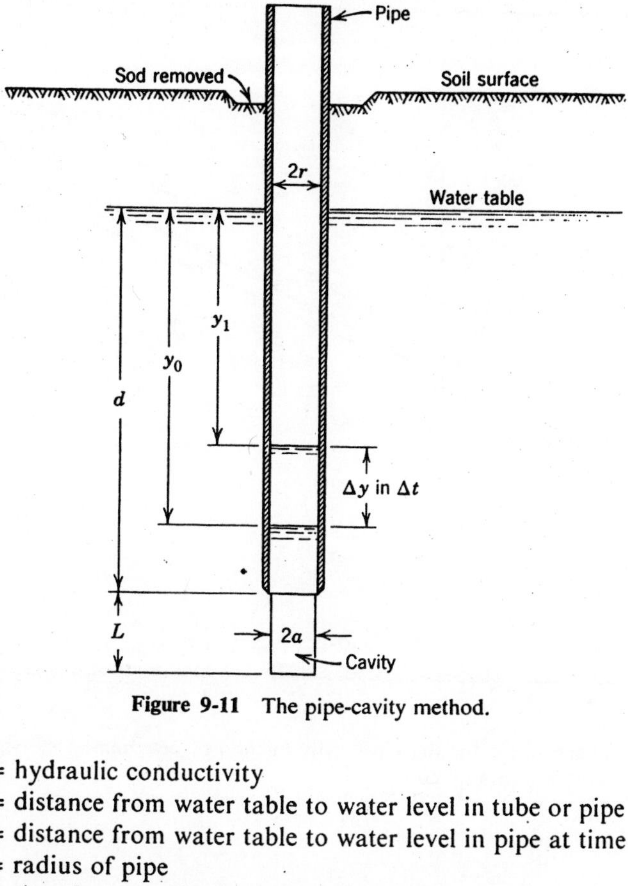

Figure 9-11 The pipe-cavity method.

where k = hydraulic conductivity
y_0 = distance from water table to water level in tube or pipe at time t_1
y_1 = distance from water table to water level in pipe at time t_2
r = radius of pipe
$t_2 - t_1$ = time for water level to change from y_0 to y_1
S = a coefficient determined with the electric analogue

The above equation is applied to the use of a pipe with or without a cavity at the end of the pipe. The field procedures of the pipe-cavity method consist of driving a pipe into an auger hole slightly smaller in diameter than the pipe according to a special technique designed to eliminate compaction.

After allowing the water table to establish itself in the pipe, a hose connected to a pump is inserted into the pipe and the pipe is pumped out. The purpose of the pumping which must sometimes be repeated a number of times is to remove puddled soil from the walls of the cavity. The inseeping water flushes out the soil pores. After puddling effects have been minimized (which may be checked by reproducibility of results on an individual hole) the soil water is allowed to rise in the pipe and the rate of rise is determined with the aid of stop watches and an electrical probe.

A simplified form of the equation 15 can be used for the actual calculation if the measurements are made while the water level in the pipe is less than half the distance to the water table (measured from the bottom of the pipe).

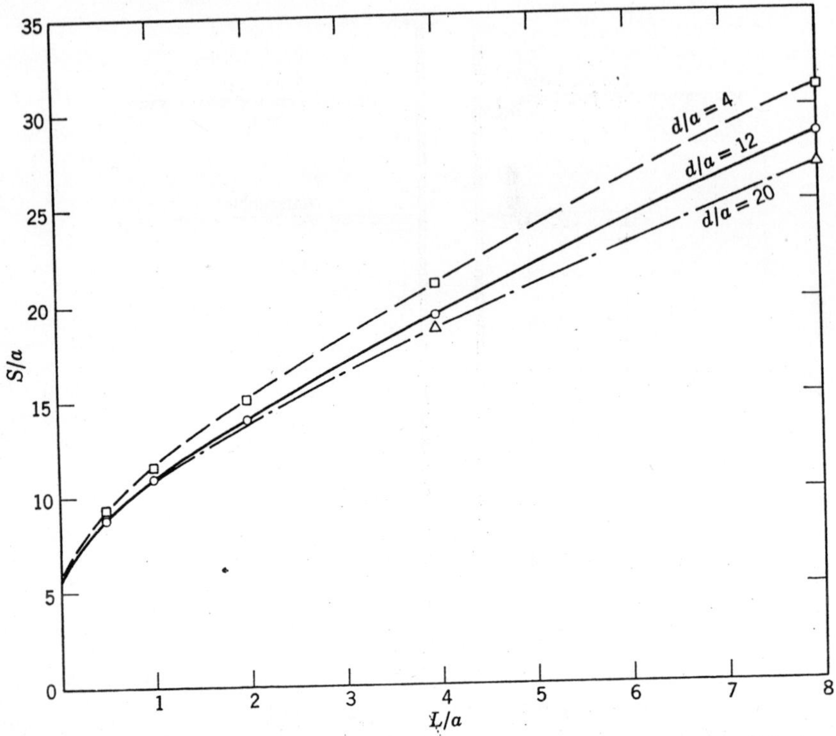

Figure 9-12 Shape factor for the pipe-cavity method of determining the soil hydraulic conductivity. After Smiles and Youngs.

Figure 9-13 Installing the pipe for measuring the soil permeability by the pipe-cavity method

Figure 9-14 Pumping water out of pipe for permeability measurement with the pipe-cavity method.

Figure 9-15 Measurement of the rise of water in the pipe.

The formula presented below is valid only for a 2-inch pipe with a cavity 4 inches long and $1\frac{15}{16}$ inches in diameter, having $S = 17$ inches.

$$k = 655 \frac{\Delta y}{\Delta t} \frac{1}{\text{average head}}$$

All measurements are made in feet and seconds, and k is given in inches per hour.

MEASUREMENT OF HYDRAULIC CONDUCTIVITY IN SITU IN THE ABSENCE OF A WATER TABLE

Shallow Well Pump-in Test

The shallow well pump-in test has been developed by the United States Bureau of Reclamation for the investigation of sites where a water table is not present. Essentially, it consists of measuring the volume of water flowing horizontally from a well in which a constant head of water is maintained by a float valve. The horizontal permeability determined by this method is a composite rate for the full depth of the hole being tested, but reflects primarily the permeability of the more permeable layers.

A hole is dug by hand to the desired depth. A float apparatus for maintaining constant head of water in the hole is installed. This float apparatus is connected by tubing to a calibrated supply tank which is on a platform beside the hole. The hole is then filled to the level of the float valve and the water level in the hole is maintained constant by means of this valve. The time and the reading on the tank gauge are recorded when everything is operating satisfactorily. A record is kept of the time and the tank-gauge reading. Water is added each time the site is visited. When water-temperature fluctuations exceed 2°C viscosity corrections should be applied.

The test should be continued until the material around the hole has become saturated and the flow from the tank is relatively constant. The permeability should be computed after each visit. When a relatively constant permeability has been reached over a 24-hour period it can be assumed that the periphery of the hole is saturated.

One of the principal limitations of this test is that it requires 2 to 6 days and a considerable amount of equipment. A relatively large amount of water is required also, especially if the material has a permeability above 2 or 3 inches per hour.

The Permeameter Method

After the proper site has been selected, a hole 4 by 4 feet is dug to within 3 inches of the layer to be tested. The last 3 inches are excavated when the equipment is ready to be installed. The equipment consists of an 18-inch cylinder. The cylinder is driven 6 inches into the soil in the middle of this large hole. About one inch of clean, uniform, permeable sand is spread over the area inside the cylinder to restrict puddling of the soil surface during the test.

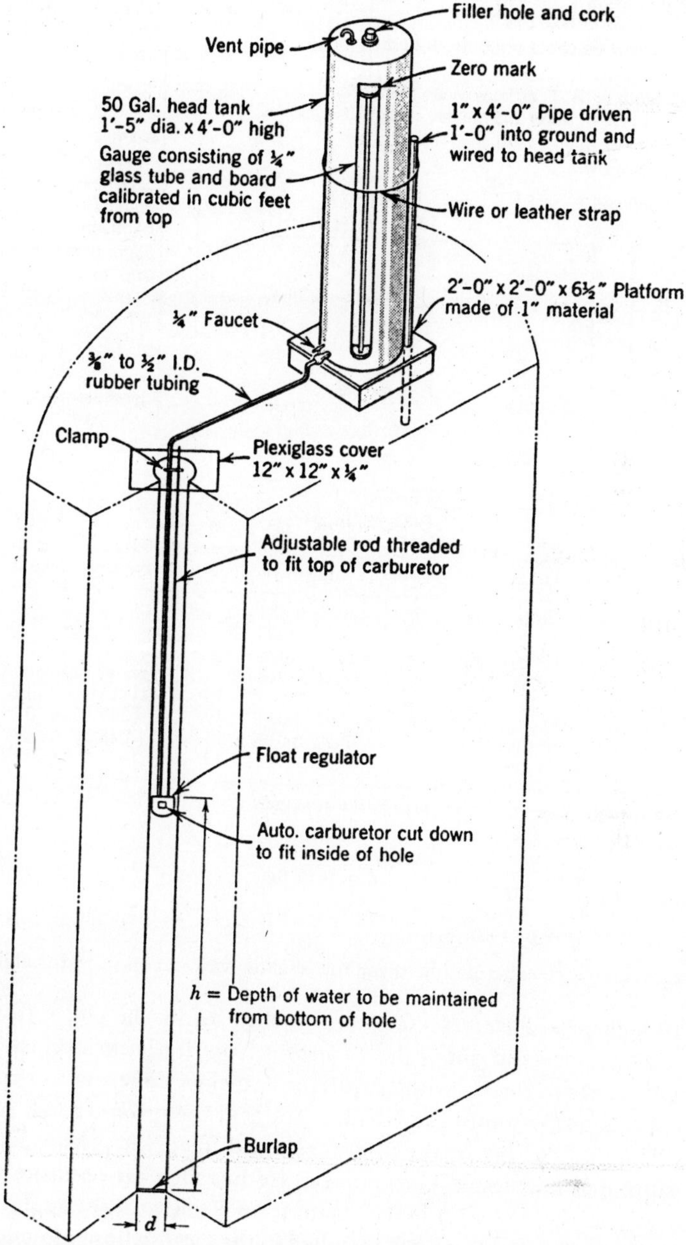

Figure 9-16 Equipment for shallow-well pump-in text. After Winger 1960.

$D = 6.0$ total depth of hole (feet)
$r = 0.167$ radius of hole (feet)
$W.T.$ or impervious strata $= 7.0$ depth below ground surface (feet)
$T_u = 4.5$ depth of $W.T.$ or impervious strata from surface of water maintained (feet)
$h = 3.5$ depth of water maintained from bottom of hole (feet)

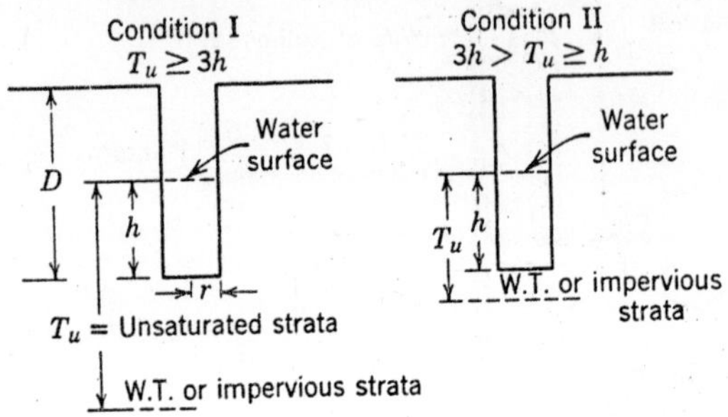

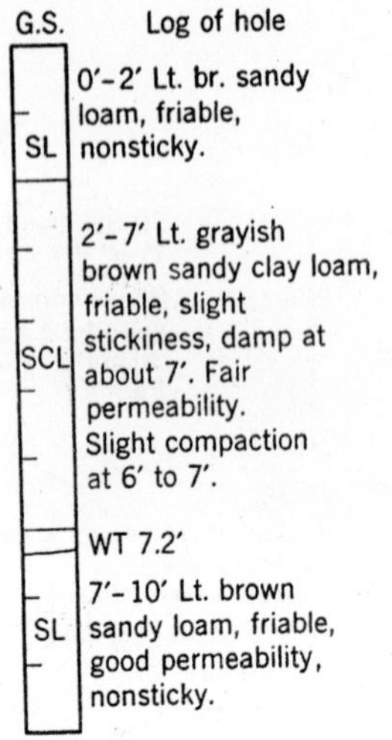

Initial Date	Initial Time	Final Date	Final Time	Time min	Tank reading Cu ft Initial	Tank reading Cu ft Final	Q cubic ft/min	Temp of water °C	Viscosity of water Centipoise	Adj Q cubic ft/min	Hydr cond in./hr
10- 8-58	0800	10- 8	1100	180	0	6.12	0.034				
10- 8-58	1100	10- 8	1400	180	0	5.97	0.033				
10- 8-58	1400	10- 8	1800	240	0	6.00	0.025	Note: Connected two barrels for greater capacity			
10- 9-58	1800	10- 9	0530	690	0	12.41	0.018				
10- 9-58	0530	10- 9	1130	360	0	6.82	0.019	16	1.1111	0.019	0.90
10- 9-58	1130	10- 9	1800	390	0	7.65	0.020	19	1.0299	0.019	0.90
10- 9-58	1800	10-10	0530	690	0	12.10	0.018	13	1.2028	0.020	0.95
10-10-58	0530	10-10	1130	360	0	6.63	0.018	15	1.1404	0.019	0.90

Remarks: No trouble with apparatus, assumed test satisfactory and results reliable.

Calculation: $h = 3.5$ (ft) $\quad r = 0.167$ (ft)

$T_u = 4.5$ (ft) $\quad h/r = 20.96$

$K = 0.90''/\text{hr}$

(Use Condition II): $\dfrac{h}{T_u} = 0.78$

$Q = 0.019\ \text{ft}^3/\text{min}$

Figure 9-17 Data and computation sheet for shallow-well pump-in permeability test.

Two 18-inch piezometers are driven 9 inches below the soil on opposite sides of the cylinder and about 3 to 4 inches from it. They are installed by driving two or three inches with the driver and then augering out the core, continuing this process until the 9-inch mark is at the ground level. A 4-inch cavity is then augered below the piezometer and filled with clean, fine sand.

Two calibrated and tested tensiometers are installed on opposite sides of the cylinder and three to four inches from it on a line at right angles to that of the piezometer. A float valve is installed in the large cylinder to maintain a constant 6-inch head. The float valve is connected to a head tank with $\frac{3}{8}$-inch rubber tubing. When the tensiometers read zero tension, no water shows in the piezometer, and water is moving through the 6-inch test layer at a constant rate, it can be assumed that the requirements of Darcy's law have been met.

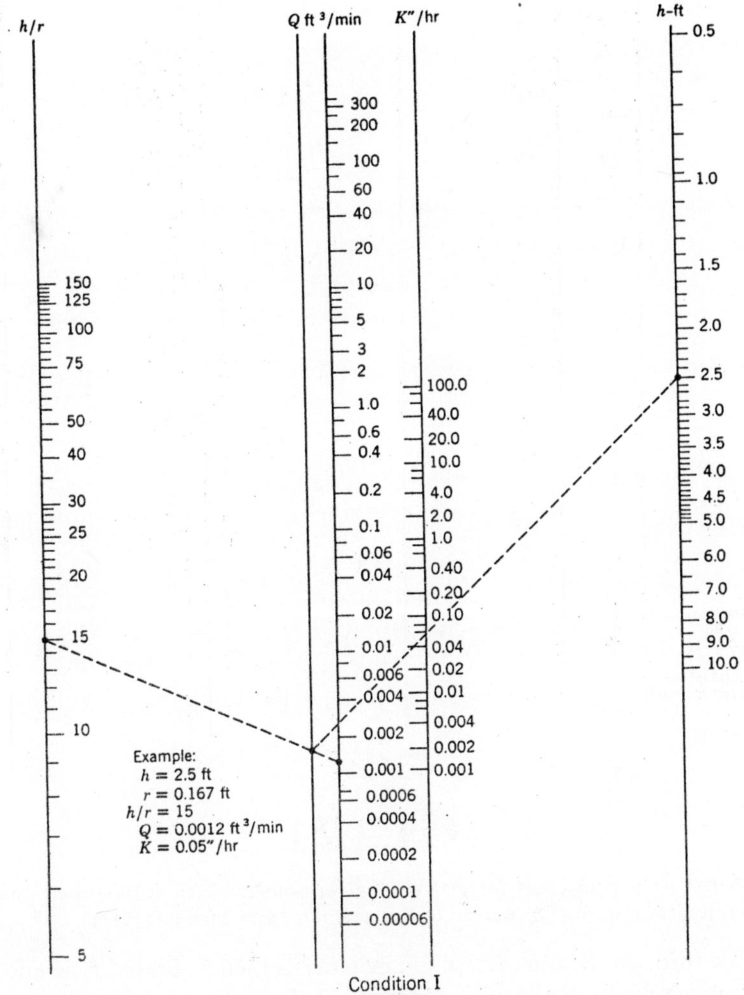

Condition I

$T_u \geqq 3h$

$$K = \frac{720[\log_e(\frac{h}{r} + \sqrt{(\frac{h}{r})^2 - 1}) - 1]Q}{2\pi h^2}$$

Figure 9-18 Alignment chart or determining permeability from shallow-well pump-in test data. Redrawn from U.S. Bureau of Reclamation Earth Manual (1951).

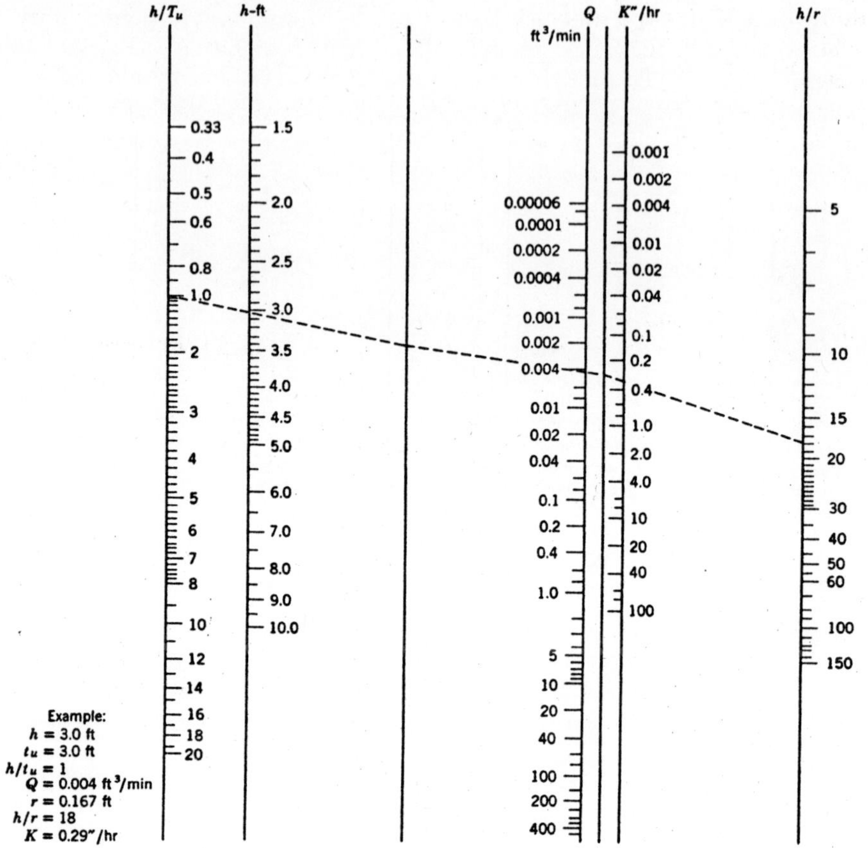

Condition II

$3h > T_u \geq h$

$$K = 720\left[\frac{3 \log_e \frac{h}{r} Q}{\pi h(h+2t_u)}\right]$$

Figure 9-19 Alignment chart for determining permeability from shallow-well pump-in test data. Redrawn from U.S. Bureau of Reclamation Earth Manual (1951).

The flow through the 6-inch test soil cylinder is then saturated flow and can be calculated by an application of Darcy's law.

Pond-Infiltration Test

In order to get away from the soil compression that is inherent in the core samples, an infiltration test over a large area has been recommended and put in practice. The area, recommended is 400 cm in diameter. The area is diked with a ring of soil and filled with water to form a pond. A circular pond is recommended rather than a rectangular one because the circular pond has less lateral and undesirable seepage loss per unit area of pond than a rectangular one. The pond test procedure is as follows: Water is added to the pond as needed. When sufficient water has been added to soak the soil down through the questionable layer, that is, the layer whose permeability is being analyzed, the falling water level of the pond in the absence of added water is observed. This rate should be a measure of the ability of the soil to pass irrigation and/or drainage water into and through the so-called impermeable layer. Since the flow in this case is almost entirely due to gravity, the hydraulic gradient will be unity and the permeability is calculated from a simple application of Darcy's law that assumes the hydraulic gradient is unity.

REFERENCES

Aronovici, V. S. 1947. The mechanical analysis as an index of subsoil permeability. *Soil Sci. Soc. Am. Proc.*, **11**:137–141.

Beers, W. J. F. van, 1958. *The Auger-Hole Method. Bull.* 1. Int. Inst. for Recl. and Imp. Wageningen, The Netherlands.

Bradshaw, G. B., and W. W. Donnan. 1950. *A falling Head Permeameter for Evaluating Permeability. U.S. Dept. Agr. Soil Cons. Serv.* Mimeograph.

Childs, E. C., and N. Collis-George. 1950. The permeability of porous materials *Proc. Roy. Soc.* (*London*), **A201**:392–405.

Fair, G. M., and Hatch, L. P. 1933. Fundamental factors governing the streamline flow of water through sand. *J. Am. Water Works Assoc.*, **25**:1551–1665.

Fireman, M. 1944. Permeability measurements on disturbed soil samples. *Soil Sci*, **58**:337–353.

Kirkham, Don. 1946. Proposed method for field measurement of permeability of soil below the water table. *Soil Sci. Soc. Am. Proc.*, **10**:58–68.

Kirkham, Don. 1955. Measurement of the hydraulic conductivity of soil in place. Symposium on Permeability of Soils. *Am. Soc. Testing Mater. Spec. Tech. Pub.*, **163**:80–97.

Kirkham, Don. 1958. Theory of seepage of water into an auger hole above an impermeable layer. *Soil Sci. Soc. Am. Proc.* **22**:204–208.

Kirkham, Don, and C. H. M. van Bavel. 1949. Theory of seepage into auger holes. *Soil. Sci. Soc. Am. Proc.*, **13**:75–82.

Kozeny, J. 1927. Ueber capillare Leitung des Wassers im Boden. *Sitzungsber. Wien. Akad. Wissensch.*, **136(2a)**:271–306.

Luthin, J. N., and D. Kirkham. 1949. A piezometer method for measuring permeability of soil *in situ* below a water table. *Soil Sci.*, **68**:349–358.

Marshall, T. J. 1957. Permeability and the size distribution of pores. *Nature*, **180**:664–665.

Maasland, M., and H. C. Haskew. 1957. The auger hole method of measuring the hydraulic conductivity of soil and its application to tile drainage problems. 3rd Cong. Intl. Comm. Irrig. and Drainage. R. 5, *Questions*, **8**:8.69–8.14.

Neal, J. H. 1934. Proper spacing and depth of tile drains determined by the physical properties of the soil. *Minnesota Agr. Exp. Sta. Tech. Bull.* 101.

Smiles, D. E., and E. G. Youngs. 1965. Hydraulic conductivity determinations by several field methods in a sand tank. Soil Sci. 99:83–87.

Wenzel, L. K. 1942. Methods for determining permeability of water bearing materials. *U.S. Geol. Surv., Water Supply Paper* 887.

Winger, R. J. 1960. *In-place Permeability Tests and their Use in Subsurface Drainage.* Int. Comm. of Irrigation and Drainage, Madrid.

PROBLEMS

1. A hole is dug in the soil for the auger-hole measurement of the hydraulic conductivity. The following information is obtained.
   $a = 3$ cm
   $y_0 = 85$ cm
   $y_1 = 82$ cm
   time = 44 seconds
   $d = 100$ cm
   Calculate k using Hooghoudt's formula, and use Ernst's graphs for the impermeable layer at the bottom of the auger hole.
2. Using a falling-head permeameter with a 3.5-mm inside diameter glass tube the following data is collected.
   Head differential from 15 cm to 5 cm
   Time of fall—10 seconds
   Temperature of water—69°F.
   What is the conductivity in cm/hour?
   The permeameter is a Type A having an inside diameter of 2 inches and length of $5\frac{1}{2}$ inches. The heads must be corrected for capillary rise.
3. A pipe having an inside diameter of $1\frac{7}{8}$ inches is driven into the soil 2 feet below the water table. A cavity 4 inches long is augered out beneath the end of the pipe. It takes 20 seconds for the water in the pipe to rise from a position 5.8 inches above the end of the pipe to a position 6.0 inches above the end of the pipe. What is the hydraulic conductivity of the soil?

Chapter 10 DEPTH AND SPACING OF DRAINS

A drainage theory describes the flow of water in an idealized soil-water system. The actual field problem is simplified to make it possible to get a mathematical solution. There is an infinite variety of soil conditions that one can find in the field and therefore drainage theories only approximate field conditions. Before applying a theory to a particular field problem it is first necessary to examine the assumptions made in its derivation. These assumptions should be compared to the problem at hand. In most cases the assumptions will not exactly correspond to the situation encountered in the field. It is then necessary to use judgment in applying the theory. In some cases the theory will work out quite well and is suitable for determining the depth and spacing of the drains. In other cases the theory is useful only as a first approximation to the proper design. In any case the drainage equations can give a great deal of aid in the solution of a problem, and they often can be applied by untrained personnel who do not have the necessary experience.

In addition to the practical aspects of the depth and spacing of drains, there are certain conclusions that can be drawn from a theoretical examination of a problem, conclusions that can be reached in no other way. These conclusions then help us to reason about problems in a rational way even though the particular theory is not directly applicable.

As stated above there is a great variety of drainage situations that occur in the field. It is necessary to select the conditions that seem to be prevalent in your area and apply the appropriate theory. The goal of the drainage system is to provide drainage for the individual farm, but in doing so it is necessary to consider the need for main drains, and other features such as the availability of water for plant growth during the summer months of low rainfall. In the analysis of problems in humid areas the Dutch have been leaders in developing techniques and methods that can be used in this analysis. Many of these methods are still in development. Only the proven techniques will be reported here. More precise techniques will undoubtedly be developed in the future but the methods given here will provide a satisfactory approximation to the problem.

In irrigated regions the Bureau of Reclamation has led in the development of the analysis of the drainage problems. Although the problems in irrigated areas are somewhat different than those encountered in humid areas there is a remarkable similarity in the methods of analysis.

One important solution of the drainage problem is a result of the work of Dr. S. B. Hooghoudt of the Netherlands. He considered the water table in equilibrium with the rainfall. The problem solved by Hooghoudt is essentially this: how high will the water table rise for a given rainfall, soil permeability, depth of drain, and spacing of drain? It is also necessary to know the depth to the barrier layer which restricts the downward flow. If drains are installed the water table will rise until the flow into the drains is just equal to the amount of rain or irrigation water infiltrating through the soil surface. At this time the water table is said to be in equilibrium with the rainfall or irrigation water. The problem is to determine the position of the water table at equilibrium. The position of the water table will depend on the following factors:

1. The rainfall rate or the rate at which the irrigation water is applied
2. The soil hydraulic conductivity
3. The depth and spacing of the drains
4. The depth to an impermeable layer

Other factors such as the rate of plant use of the water, deep seepage, soil stratification, and so forth are usually ignored in the analysis in order to simplify the mathematical treatment.

The above assumptions are then incorporated into a mathematical analysis of the problem that gives as its result the height to which the water table will rise under a specific set of conditions. After determining the height of the water table it is necessary to know whether or not injury will be caused to the plants or to the soil.

A great many field observations and measurements have been made in Holland to determine the important aspect of control of the water table for maximum plant growth. There are two separate factors which are of equal importance in the determination. In the first place it is necessary to keep the water table low enough during the winter months so that the growth of winter crops is not restricted. It is during the winter months that most of the rain falls in Holland, and the drains provide for the lowering of the winter water table. The depth varies with the soil and with the crop that is grown.

During the summer months there is a rainfall deficiency, especially on the lighter soils. During this time it is required that the water table be kept high to supply the plant needs. During the summer months the flow towards the drains is reversed and the flow occurs from the drains out into the field.

Other approaches to the development of drainage theory will be considered later in the chapter. The steady-state problem outlined above is one that has been used successfully by Dr. S. B. Hooghoudt of Holland as well as Dr. Don Kirkham of the United States in developing usable drainage theories. Dr. Hooghoudt's theory will be considered first. It is based on some simplifying assumptions that Kirkham was able to avoid. However, Hooghoudt's theory requires a knowledge of elementary mathematics and has the added advantage of a rather simple solution, whereas Kirkham's analysis utilizes advanced mathematics and the solution is expressed in terms of complicated mathematical functions.

HOOGHOUDT'S EQUATION FOR THE WATER TABLE IN EQUILIBRIUM WITH RAINFALL OR IRRIGATION WATER

The problem analyzed by Hooghoudt is presented in Figure 10-1 which shows a homogeneous soil of known permeability with an impermeable stratum lying under it. The soil is assumed to be drained by a series of parallel ditches. It will be shown that the same analysis can be applied to subsurface drains as well.

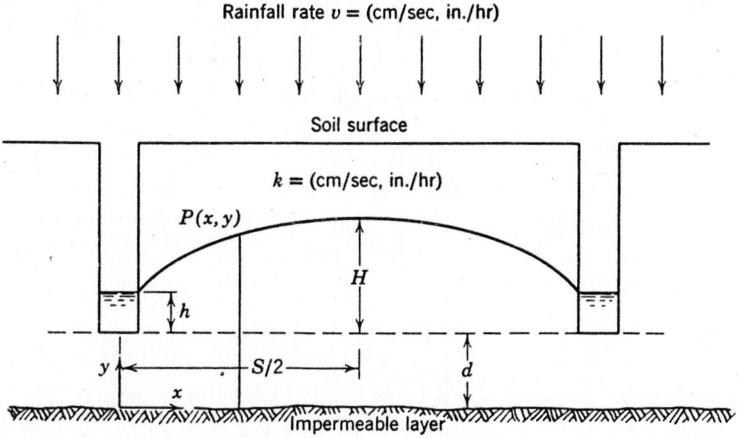

Figure 10-1 Diagram for Hooghoudt's drain-spacing formula. The water table is in equilibrium with the rainfall or irrigation water.

In Hooghoudt's analysis it is assumed that rain is falling at a constant rate on the soil surface. In order to simplify the mathematical analysis, it is assumed that the hydraulic gradient at any point is equal to the slope of the water table above that point. This assumption is known as the Dupuit-Forchheimer (D-F) assumption.

The D-F assumptions imply that water flows horizontally because all the equipotentials are vertical planes. This is, of course, an erroneous picture of the actual flow paths of the water. It is especially incorrect near the drains where the flow paths are quite curved. However, where the slope of the water table is relatively flat the D-F assumptions are nearly valid. The strength of the D-F assumptions lies in the fact that the resulting equations give an accurate value (within 10% of the true value) for the total flow into the drainage facility, even though the individual flow paths are not described accurately.

Hooghoudt's assumptions can be summarized as follows:

1. The soil is homogeneous and of hydraulic conductivity k.
2. The drains are evenly spaced a distance S apart.

3. The hydraulic gradient at any point is equal to the slope of the water table above the point, dy/dx.
4. Darcy's law is valid for flow of water through soils.
5. An impermeable layer underlies the drain at a depth d.
6. Rain is falling or irrigation water is applied at a rate v.
7. The origin of coordinates is taken on the impermeable layer below the center of one of the drains.

It is evident from an examination of Figure 10-1 that a vertical plane drawn between the center of the two drains is a division plane for the water. All the water entering the soil to the right of this plane flows into the right drain and, similarly, all of the water on the left goes to the left drain.

First consider the flow through a vertical plane drawn from the point P on the water table to the impermeable layer. All the water entering the soil to the right of this plane must pass through it on its way to the drain. Since v is the quantity of water entering a unit area of the soil surface then the total quantity of water passing through the plane will be equal to v multiplied by the surface area from the plane to the midpoint between the tile lines. The surface area is equal to $(S/2 - x) \cdot 1$, where 1 stands for a unit distance measured out from the paper. In other words we consider a unit thickness of soil. The quantity of water flowing per unit time through the plane is given by,

$$q_x = \left(\frac{S}{2} - x\right) v \tag{1}$$

We can obtain a second expression for q_x by applying Darcy's law to the flow through the plane. First, remember that the hydraulic gradient at any point is assumed to be equal to the slope of the water table above the point. In other words the hydraulic gradient is equal to dy/dx. Since the distance from the impermeable layer to the water table is y, the cross sectional area of flow at the plane is equal to y. Substituting these values in Darcy's law gives

$$q_x = ky \frac{dy}{dx} \tag{2}$$

The right side of equation 1 must equal the right side of equation 2 since the flow in the two instances must be equal. Therefore

$$\left(\frac{S}{2} - x\right) v = ky \frac{dy}{dx}$$

Multiplying through by dx gives

$$\left(\frac{S}{2} - x\right) v \, dx = ky \, dy$$

or

$$\frac{vS}{2} dx - vx \, dx = ky \, dy$$

This is an ordinary differential equation and can be integrated as follows:

$$\int \frac{vS}{2}\,dx - \int vx\,dx = \int ky\,dy$$

On integration

$$\left(\frac{vS}{2}\right)x - \frac{vx^2}{2} = \frac{ky^2}{2}$$

The limits of integration are $x = 0$ then $y = h + d$, and when $x = S/2$ then $y = H + d$. Substituting these limits we have

$$(vS/2)x\Big]_0^{S/2} - \tfrac{1}{2}vx^2\Big]_0^{S/2} = \tfrac{1}{2}ky^2\Big]_{h+d}^{H+d}$$

which results in

$$S^2 = \frac{4k(H^2 - h^2 + 2\,dH - 2\,dh)}{v}$$

which is Hooghoudt's equation for either open ditch drains or subsurface drains such as tile drains.

An important point which will be discussed later is the factor d, the distance from the bottom of the drain to the impermeable layer. As it goes to infinity so does S, the drain spacing. This is because the D-F assumptions do not properly account for the radial flow into the bottom of the drain.

For practical purposes the drain is considered to be empty. Hooghoudt's equation then reduces to

$$S^2 = \frac{4kH}{v}(2d + H) \qquad (3)$$

It is this equation that has been used in Holland (Van Someren), Australia (Maasland, 1956), and in the United States (Donnan, Aronovici, Blaney, 1947) for design purposes.

Equation 3 is the equation of an ellipse. This can be seen by transforming the origin of the coordinate system to the midpoint between the drains. The equation as written before the substitution of limits is

$$\frac{vS}{2}x - \frac{vx^2}{2} = \frac{ky^2}{2}$$

To move the origin of the coordinate system to the midpoint between the drains we substitute the transformation

$$x_1 = \frac{S}{2} - x$$

On substitution we obtain

$$\frac{y^2}{S^2v/4k} + \frac{x^2}{S^2/4} = 1$$

which is the equation of an ellipse having semimajor and semiminor axes given by $S/2$ and $S/2\sqrt{v/k}$ respectively.

HOOGHOUDT'S EQUATION FOR A LAYERED SOIL

In the event that a soil consists of two layers of different hydraulic conductivity then it is possible to use Hooghoudt's procedures to derive a drain spacing formula.

If k_a is the hydraulic conductivity of the layer above the drain line and k_b the hydraulic conductivity below the drain line then Hooghoudt's formula becomes

$$S^2 = \frac{4}{v}(k_a H^2) + \left(\frac{8}{v} k_b\, dH\right)$$

where d is the equivalent depth obtained from Hooghoudt's graphs. A multilayered soil can be treated by taking a weighted-mean of the horizontal conductivities. For example suppose the layer above the drain line consists of three layers of conductivity k_1, k_2, and k_3 having thickness l_1, l_2, l_3. The average will be

$$k_a = \frac{k_1 l_1 + k_2 l_2 + k_3 l_3}{l_1 + l_2 + l_3}$$

THE USE OF HOOGHOUDT'S EQUATION

Depth to Impermeable Layer

As stated above the formula is not valid for large values of d, the depth to the impermeable layer. Hooghoudt recognized this difficulty and made a separate analysis for the flow beneath the drain. He assumed that the flow

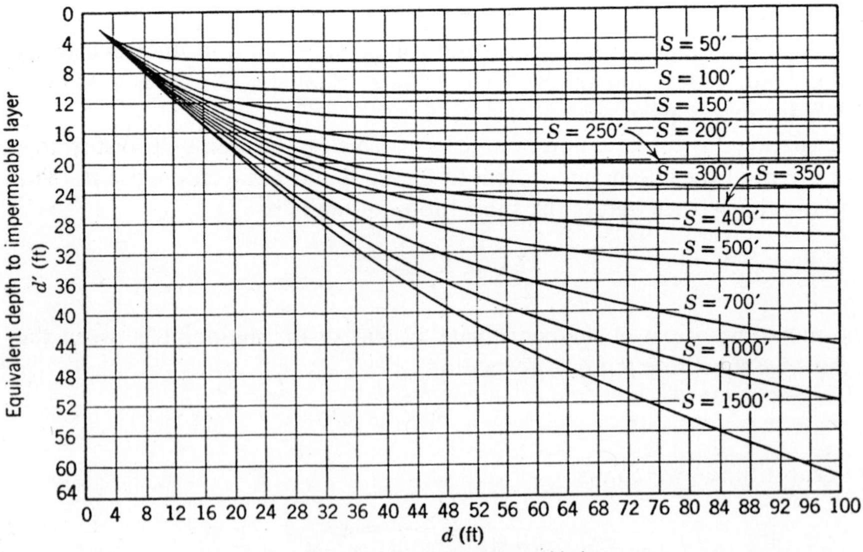

Figure 10-2 Relationship between d and d' where $r = 0.7$ ft and S is the spacing between the drains. Curves are based on Hooghoudt's correction (after Bureau of Reclamation).

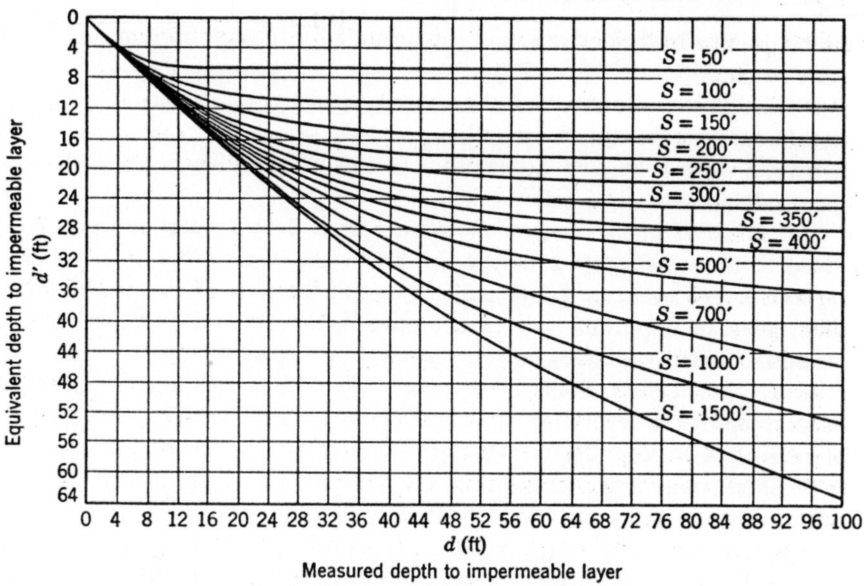

Figure 10-3 Relationship between d and d' where $r = 0.8$ ft and S is the spacing between the drains. Curves based on Hooghoudt's correction (after Bureau of Reclamation).

is radial in character. He then compared the flow obtained with the radial-flow assumptions to the flow obtained with the horizontal-flow equation and developed a table of "equivalent" depths. Wesseling (1964) indicates that Hooghoudt's table of equivalent depths is correct to about 5%. The values of equivalent depths obtained by Hooghoudt are to be substituted in equation 3 for d.

The determination of the impermeable layer is often troublesome in the field. How impermeable does the layer have to be before it is considered impermeable? If the underlying layer has one-tenth the permeability of the upper layer then it can be considered impermeable from the standpoint of analysis. This does not mean that water cannot or does not seep down through the so-called impermeable layer. Significant amounts of water can be expected to be the lost through this layer, although the flow pattern is not seriously affected by this flow.

DETERMINATION OF THE RAINFALL OR IRRIGATION RATE. The rate at which water moves into the water-table zone is actually determined by the infiltration rate of the soil. However, in the use of Hooghoudt's equation the value of v is based on the rainfall rate or the irrigation rate. In rainfall areas the rainfall is seldom if ever constant over long periods of time. In actual fact the value of v depends on many other factors besides the rate of rainfall or the rate of application of irrigation water. Consider first the situation in humid regions where the drainage problem is caused by excess rainfall.

A more accurate description of v would be to call it the rate at which water replenishes the zone beneath the water table. Such a description is implicit in the use of v in the derivation of the formula for the equilibrium or steady-state situation.

Obviously the rate at which the water replenishes the water-table zone is not the same as the rate of rainfall, nor is it the same as the rate at which the irrigation water is applied. Some of the factors which influence the replenishment rate are described below.

Soil Moisture at the Time of the Rain Storm. Before water can percolate down to the water-table zone the soil above the water table must become saturated. The moisture content of the soil at the start of the precipitation thus plays an important role in determining the amount and the time at which the water percolates into the groundwater zone.

Interception Losses. Some of the water which falls on the soil surface is intercepted by the vegetation and never reaches the soil surface. The amount of interception depends on the type of crop. A tree crop can intercept substantial amounts of precipitation. The amount of interception depends to a certain extent on the rate at which the precipitation is falling.

Deep Seepage. At the same time the precipitation is occurring there may be seepage through layers lying beneath the drainage level of the drains. The deep seepage must be subtracted from the rainfall to obtain the amount of replenishment.

Artesian Seepage. Upward movement of water from deeper aquifers may add to the amount of water to be drained.

Surface Runoff. The amount of surface runoff will depend on the soil-infiltration rate, the slope of the land surface, and the rate at which the precipitation is falling. In addition, the soil moisture content at the time of the precipitation has an influence on the infiltration and hence the surface runoff.

Evapotranspiration. Some of the water which falls will be evaporated from the soil surface and some will be transpired by the plants.

In conclusion we can say that v involves a complex of hydrologic factors. The best way to obtain a value of v is to measure the outflow of existing drains. Surface runoff must be excluded from these measurements.

In areas being drained for the first time and where there are no existing drains on which the measurements can be made it is recommended that values for similar areas be obtained from the literature. After the initial installation of drains some measurements should be made to check the accuracy of the assumed value of v.

APPLICATION OF HOOGHOUDT'S EQUATION IN HUMID AREAS

In Holland the normal criterion states that with a discharge of 0.007 m/day (7mm/day) the water table may not be higher than 50 cm below the soil surface for arable land and 40 cm below the surface for grassland. Under these conditions the drain depth is about 80 to 90 cm.

One approach used by the Dutch which takes into account the rainfall intensity is as follows. The drainage rate, or v, is taken to be equal to about

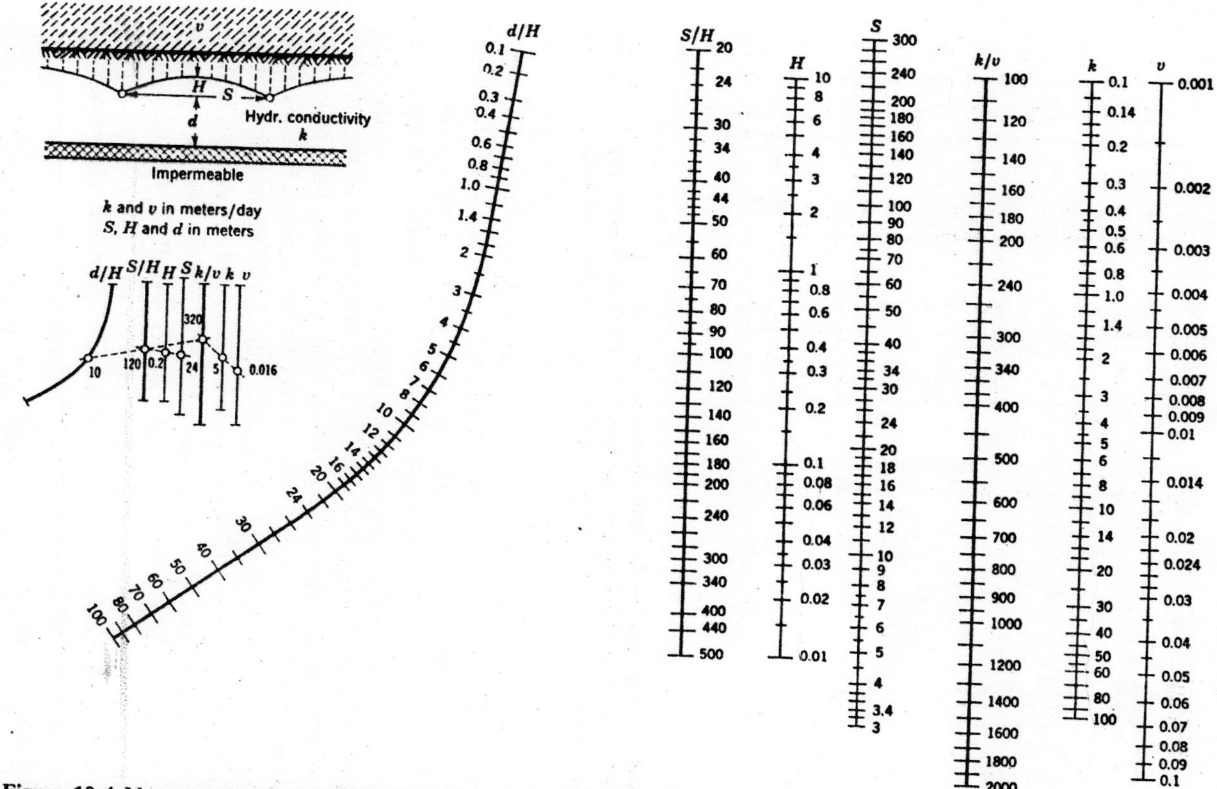

Figure 10-4 Nomograph for the determination of drain spacings with $k/v \geq 100$. From Visser (1954, p. 77). With permission from Research Division, Netherlands Government Service for Land and Water Use. Original drawing by J. H. Boumans from data of L. F. Ernst.

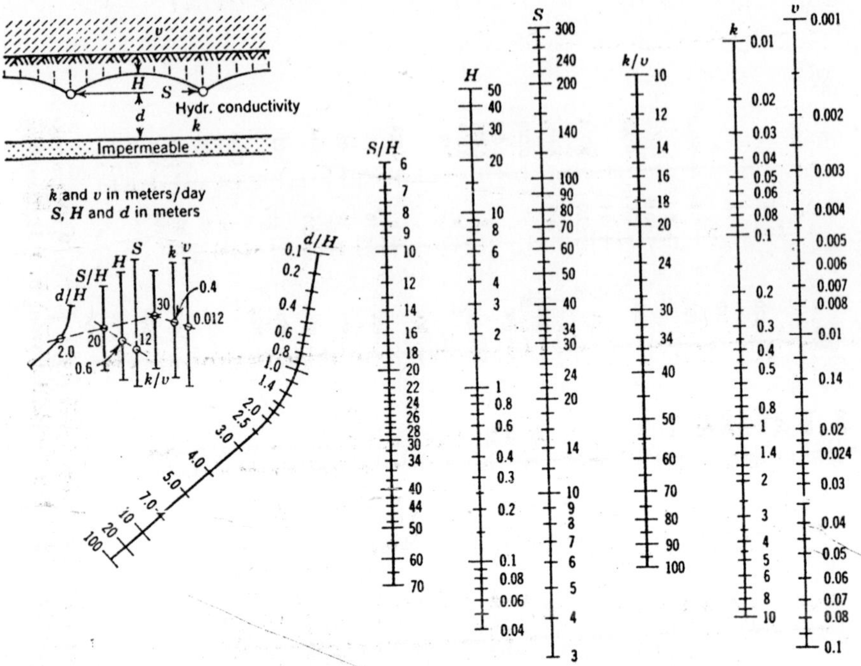

Figure 10-5 Nomograph for the determination of drain spacings with $k/v \leq 100$. From Visser (1954, p. 77). With permission from Research Division, Netherlands Government Service for Land and Water Use. Original drawing by J. H. Boumans from data of L. F. Ernst.

5 to 7 mm/day. A surface runoff of about 15 mm/day is assumed. The height to which the water table will rise is then plotted as a function of the rainfall rate. The frequency of the rainfall rate is indicated on the diagram. For example, the height of the water table is plotted for rainfalls which are known to occur in two days, three days, and so forth. Then, from a study of the rainfall distribution through the year, and a consideration of the tolerance of the plant to high water-table conditions, it is decided how serious the drainage problem is. The time of the year is taken into account by considering the temperature, which is held to be a good indicator of the damage that will result to the plant from high water-table conditions. It is thought that the number of day-degrees that the temperature is above zero centigrade is directly related to the damage that will result to the plant. It is recognized that this relationship is not completely true since some plants, such as tomatoes, do not suffer from high water-table conditions until the temperature is about 6 or 7 degrees above zero.

The equation to use is

$$t \times T^{\circ} = 30$$

If a value of 30 is exceeded, the plants will suffer from serious damage and additional drainage must be provided.

USE OF HOOGHOUDT'S EQUATION IN IRRIGATED AREAS

In irrigated areas the problem of determining v is similar. Donnan and his colleagues have made extensive application of Hooghoudt's equation to the drainage of irrigated areas. They point out the necessity of determining the amount of water to be drained out of the soil to satisfy the leaching requirement and to keep the water table at a safe level.

The Leaching Requirement

The leaching requirement is defined as the fraction of the irrigation water that must be leached through the root zone of the plants in order to prevent the soil salinity from exceeding a specified level. The leaching requirement depends on the salinity of the applied irrigation water as well as the maximum concentration of salts that can be permitted in the root zone of the plants.

The concept of the leaching requirement is very useful in computing the amounts of drainage water that must be removed from a large area. However, such a computation presupposes that there is adequate control of the applied irrigation water and that the applied irrigation water is just equal to the crop needs plus the leaching requirement.

The maximum concentration of salts, with the exception of surface crusts formed by evaporation, will form at the bottom of the root zone. This concentration will be the same as the concentration of the salts in the drainage water, provided there is no excess leaching and the irrigation water is applied uniformly over the area.

The increase in concentration of salts in the drainage water over the concentration in the irrigation water is a consequence of the consumptive use of water by the crop. The crop will extract the water from the soil but will leave most of the salt behind. To summarize, the following assumptions are inherent in a simple application of the leaching requirement.

1. Uniform areal application of irrigation water
2. No rainfall
3. No removal of salt in harvested crop
4. No precipitation of salt in the soil

The calculation is based on the total equivalent depths of water used over a period of time. These simplifying assumptions ignore the moisture and salt storage in the soil, cation exchange reactions, depth of root zone, and crop use of salt. However, the concept has proved very useful.

The leaching requirement as defined earlier is equal to the ratio of the equivalent depth of drainage water to the depth of irrigation water, $D_{\mathrm{dw}}/D_{\mathrm{iw}}$. It may be expressed as a fraction or as a per cent. Alternatively, it may be expressed in terms of the electrical conductivity of the drainage water compared to the irrigation water. In this case the leaching requirement LR is given by

$$\mathrm{LR} = \frac{D_{\mathrm{dw}}}{D_{\mathrm{iw}}} = \frac{EC_{\mathrm{iw}}}{EC_{\mathrm{dw}}} \tag{4}$$

For some field crops an EC_{dw} of 8 mmhos/cm can be tolerated. For irrigation waters with conductivities of 1, 2, and 3 mmhos/cm, respectively, the leaching requirements will be 13, 25, and 38%. These figures are conservative because some of the salt is removed by the crop, and some of the salt may be precipitated in the form of salts such as calcium carbonate or gypsum.

In using the leaching requirement it should be borne in mind that the winter precipitation may well be adequate to leach the soil. All the water that passes through the root zone of the plant must be considered in the use of the equation. The conductivity of the irrigation water should be the weighted average of the conductivities of the rain water EC_{rw}, and the irrigation water, EC_{iw}, as shown in the equation

$$EC_{(rw+iw)} = \frac{D_{rw}EC_{rw} + D_{iw}EC_{iw}}{D_{rw} + D_{iw}}$$

where D_{rw} and D_{iw} are the depths, respectively, of the rainwater and the irrigation water that enters the soil.

In order to use the leaching-requirement concept to analyze the drainage water situation over a large area it is first necessary to know the consumptive use of the crop or crops to be grown. The amount of irrigation water will equal the sum of the consumptive use and the drainage water as given by the following equation.

$$D_{iw} = D_{cw} + D_{dw} \tag{5}$$

We can eliminate D_{dw} by using equation 4. We then have the depth of irrigation water expressed in terms of the consumptive use and the leaching requirement

$$D_{iw} = \frac{D_{cw}}{(1 - LR)}$$

Rewriting the equation in terms of the conductivity ratio we get

$$D_{iw} = \left(\frac{EC_{dw}}{EC_{dw} - EC_{iw}}\right) D_{cw}$$

It should be remembered that the EC_{dw} represents the salt tolerance of the crop to be grown.

The tolerance of some plants to salts is given in Table 10-1.

The following is an example of the use of the leaching requirement to calculate the rate of replenishment in an irrigated area.

The EC_{iw} is about 1 mmho/cm. The EC_{dw} can be taken as 8 mmho/cm and the consumptive use is 0.35 in./day

$$D_{iw} = \left(\frac{EC_{dw}}{EC_{dw} - EC_{iw}}\right) D_{cw}$$

$$= \left(\frac{8}{8 - 1}\right) 0.35 = 0.40 \text{ in./day}$$

$$v = D_{dw} = \left(\frac{1}{8}\right)(0.40) = 0.050 \text{ in./day.}$$

TABLE 10-1 Relation Between Crop Response and Soil Salinity as Determined by the Saturation Extract Method

Conductivity of Extract, mmhos/cm *at* 25°C	*Crop Response*
0–2	Salinity effects negligible for most crops
2–4	Yields of very sensitive crops may be restricted
4–8	Yields of many crops restricted
8–16	Only tolerant crops yield satisfactorily
above 16	Only a few very tolerant crops yield satisfactorily

Other Factors Influencing Drain Depth and Spacing

The effect of water table on plant growth depends on such factors as the stage of the plant growth, the climatic conditions, the soil, and the rooting habits of the plants. It is necessary to strike some sort of an average of plant conditions. From experience it appears that a water table depth in irrigated areas of about 3 feet or more is adequate to prevent salt accumulation at the soil surface and is adequate for plant growth.

Donnan and his colleagues, as a result of extensive experimentation in the Imperial Valley of California, list the following factors as having an influence on v.

1. Irrigation head
2. Length of run
3. Slope
4. Soil type
5. Infiltration rate
6. Evapotranspiration
7. Seepage into or out of the area
8. Artesian pressure
9. Type of crop-rooting habits—depth of rooting
10. Frequency of irrigation
11. Soil moisture control at the time of irrigation.

In the Imperial Valley alfalfa is the largest user of water on a yearly basis. Donnan, Aronovici, and Blaney (1947) have used it for the purpose of drainage design. They consider that 10% of the water applied must be dealt with in the design of a drainage system in the Imperial Valley. Part of this 10% may drain out of the area by deep seepage or by lateral seepage. On the other hand foreign seepage and/or artesian seepage may augment to 10%. They suggest that the technician make an inventory and review all available data concerning the problem area before estimating the excess per cent to be used in the design.

For alfalfa they suggest that 0.0036 gallons/square foot/day be drained for each 2% excess.

This is based on the following assumptions: (1) an average duty of water for alfalfa of 4.8 acre-feet per year, (2) a surface waste of 10%, (3) 12 irrigations annually, and (4) a 15-day irrigation cycle.

Donnan et al. point out that if the farmer has records of the water use, it may be advantageous to accept either his annual amount applied in lieu of the valley average, or his peak irrigation-application during the early summer months, and his cycle of days in apportioning the daily amount of water to be drained by the system.

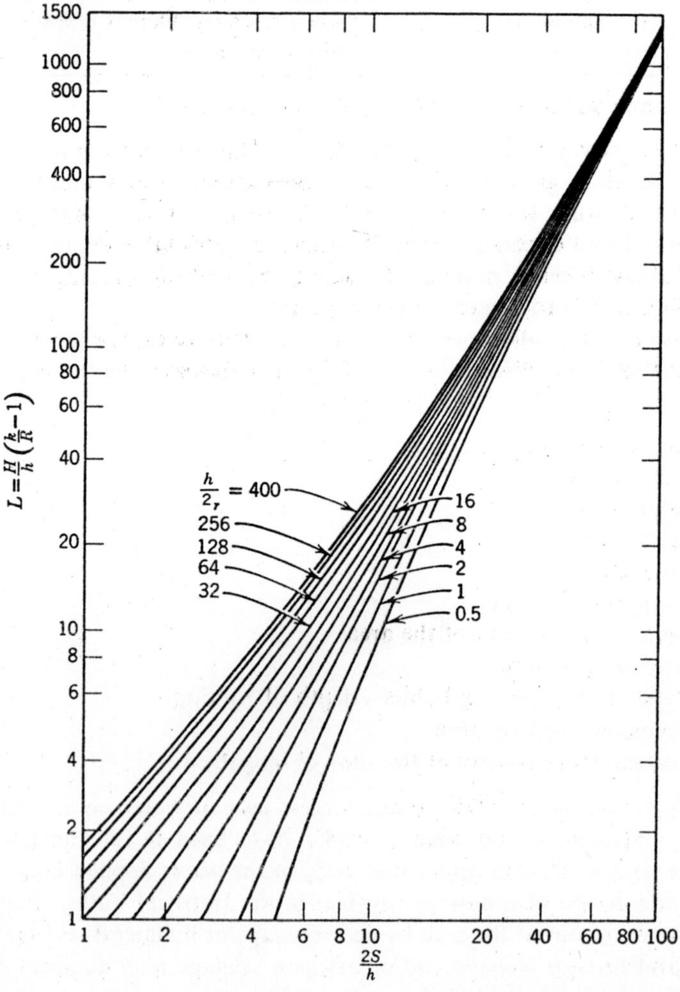

Figure 10-6 Graph for solution of Kirkham's 1958 formula (after Sadik Toksöz).

KIRKHAM'S 1958 FORMULA

Kirkham has analyzed the problem by using exact mathematical procedures. His results are, therefore, more accurate than Hooghoudt's but also they are much more complicated. Wesseling indicates that the two equations differ by less than 5%. His formula is

$$H^d = (2SR/k)F(2r/2S, h/2S)$$

where H^d = maximum height of the water table above the drains
R = rate of rainfall = v
k = hydraulic conductivity
h = distance from impermeable layer to water table immediately over drains
$2S$ = spacing of drains
r = radius of drain

where

$$F = \frac{1}{\pi}\left\{\ln\frac{2S}{\pi r} + \sum_{m=1}^{\infty}\left[\frac{1}{m}\left(\cos\frac{m\pi r}{S} - \cos m\pi\right)\left(\coth\frac{m\pi h}{S} - 1\right)\right]\right\}$$

Graphs have been prepared by Sadik Toksöz for the solution of the equation.

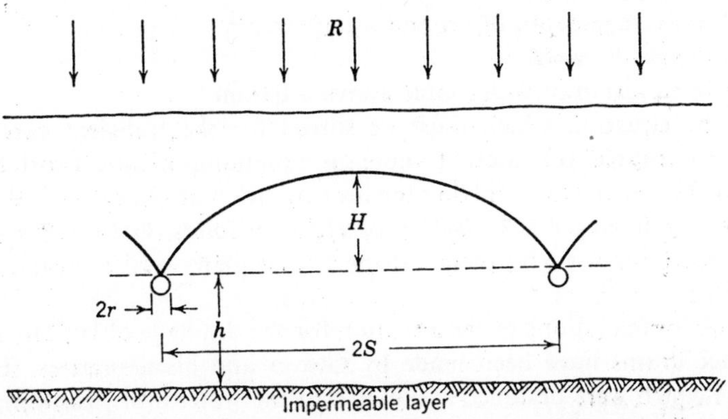

Figure 10-7 Example of the use of the graphs of Sadik Toksöz for the solution of Kirkham's 1958 equation.

$H = 0.6$ meter
$h = 6.0$ meters
$k = 1.20$ meters/day
$R = 0.20$ liter/sec/hectare = 0.00173 meter/day
$2r = 10$ centimeters = 0.10 meter

$$L = \frac{H}{h}\left(\frac{k}{R} - 1\right) = \frac{0.6}{6}\left(\frac{1.20}{0.00173} - 1\right) = 69.3$$

$$\frac{h}{2r} = \frac{6}{0.10} = 60$$

From the curve

$$\frac{2S}{h} = \frac{2S}{6} = 19.7$$

$2S = 118.2$ meters

BUREAU OF RECLAMATION FORMULA

Many designers of drainage systems have felt that the formulas which describe the water table in equilibrium with the rainfall or irrigation water do not conform to the situation in the field where the water table is constantly moving through the soil. These investigators feel that an equation that describes the movement of the water table through the soil will provide a more accurate method of design. The movement of the water table through the soil is a transient condition. The hydraulic head at any point in the soil is not constant; it is changing with time. Therefore, the situation is called the "transient" state as opposed to the "steady state" in which the hydraulic head does not vary with time.

The leaders in the study of the transient state of drainage have been the United States Bureau of Reclamation, and in particular R. D. Glover of their staff. The formulas presented below have been developed by Glover and others working for the Bureau of Reclamation. For convenience the formulas will be called the Bureau of Reclamation Formulas.

$$\frac{kD}{s}\frac{dy}{dt} = \frac{d^2y}{dx^2}$$

where k = hydraulic conductivity
D = average depth of ground-water stream
s = specific yield
y = elevation of water table above a datum

This is the equation which must be solved for the transient case of the moving water table. It is useful to note the equation is identical with the heat equation. The methods of solving the heat equation are extensively discussed in Carslaw and Jaeger (*Conduction of Heat in Solids*, Oxford Press, 1959) and the methods used by them can be used equally well for the transient water table.

A number of solutions of the equation for the drainage of land by parallel subsurface drains have been made by Glover and his associates. The first solutions which were obtained assumed that the water table was initially flat and parallel with the soil surface. Subsequent solutions were obtained by assuming other shapes for the water table. The solution that is currently recommended for use by the Bureau of Reclamation is a solution which is based on the assumption that the water table initially has a shape that corresponds with a fourth degree parabola. At time $t = 0$ the water table has a shape given by the equation

$$y = \frac{8H}{L^4}(L^3x - 3L^2x^2 + 4Lx^3 - 2x^4)$$

At the two drains the water table is taken to be at the same elevation as the drains or

$$y = 0 \qquad t = 0 \qquad x = 0$$
$$y = 0 \qquad t = 0 \qquad x = L$$

In the development, L is the spacing between the drains.

The solution to the equation for these boundary and initial conditions is

$$y = \frac{192H}{\pi^5} \sum_{m=0}^{\infty} \frac{(2m+1)^2\pi^2 - 8}{(2m+1)^5} \exp\left(-\frac{(2m+1)^2\pi\alpha t}{L^2}\right) \sin\frac{(2m+1)\pi x}{L}$$

where $\alpha = kD/s$
k = hydraulic conductivity
D = average depth of flow region
s = specific yield (per cent by volume)
L = drain spacing
H = water-table height above drain at midpoint.

Since the main interest is in the height of the water table at the midpoint between the drains, we can obtain the following expression for y as $x = L/2$:

$$H = \frac{192}{\pi^3} \sum_{n=1,3,5}^{\infty} (-1)^{(n-1)/2} \frac{n^2 - 8/\pi^2}{n^5} \exp\left(-\frac{\pi^2 n^2 \alpha t}{L^2}\right)$$

An approximate solution can be obtained by taking only the first term of the series. The Bureau of Reclamation indicates that the spacing obtained with this formula is very little different from the spacing obtained with the formula that was based on an initially flat water table. The bureau feels, however, that the use of the fourth-degree parabola to represent the initial position of the water table comes closer to the true situation.

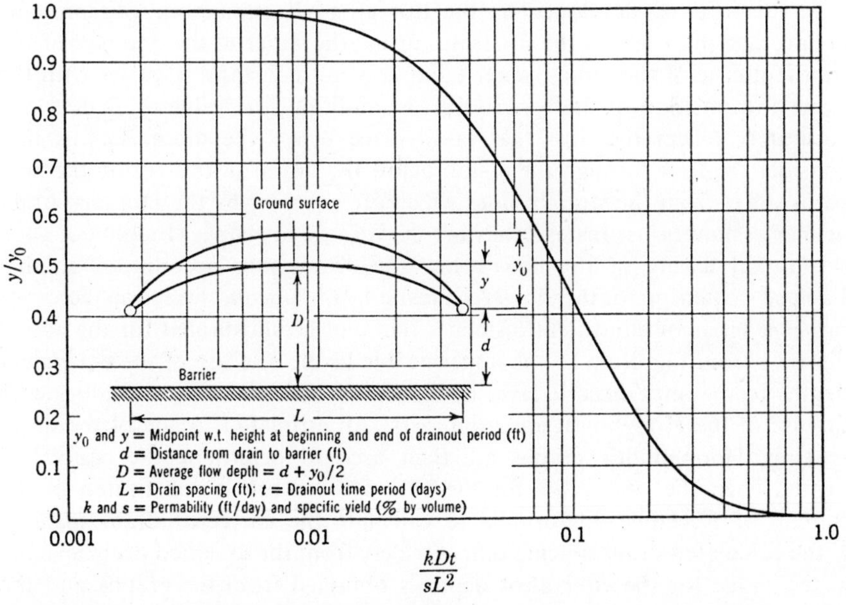

Figure 10-8 Curve showing relationship between y/y_0 and kDt/sL^2 at midpoint between drains (where drain is not on barrier).

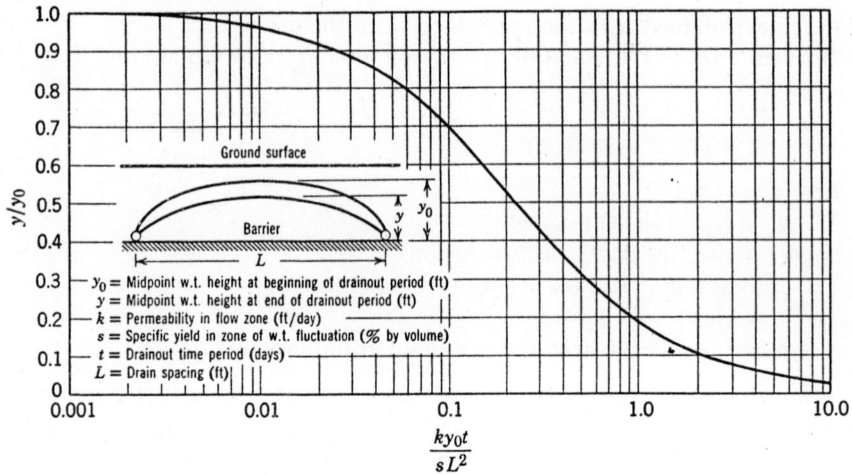

Figure 10-9 Curve showing relationship between y/y_0 and ky_0t/sL^2 at midpoint between drains (where drain is on barrier).

Curves have been prepared to show the relationship between the dimensionless parameters y/y_0 and kDt/sL^2.

THE USE OF THE BUREAU OF RECLAMATION EQUATION

The Value of D to Be Used in the Formula

In the formula developed by the Bureau of Reclamation, D represents the average thickness of the soil transmitting the water to the drains. For the transient case of the falling water table D is not constant. It varies with the slope and position of the water table. In addition, the value of D does not accurately reflect the flow that takes place below the drain. Let us first consider the flow through the region below the drain. In this region the flow paths differ from the straight lines which are required by the Dupuit-Forchheimer assumptions. Instead they are curvilinear in shape. Hooghoudt took this radial nature of the flow below the drains into account by solving Laplace's equation for this flow. As a result he obtained a correction factor, or rather a series of equivalent distances that can be substituted for the actual distance from the drain to the impermeable layer. The use of the equivalent depths to the impermeable layer is discussed under the section dealing with the use of the Hooghoudt equation. First, an estimate is made of the drain spacing. Hooghoudt's graphs are then used to determine the equivalent depth. Then the new value for the equivalent depth is substituted in the Bureau of Reclamation formula to calculate the corrected drain spacing. If the calculated drain spacing differs widely from the assumed drain spacing a new value for the equivalent depth is obtained from the graphs and the calculation repeated. It is usually necessary to repeat the calculation only once in order to obtain a satisfactory value for the equivalent depth.

Another approximation is necessary to obtain the average distance from the water table to a plane running through the drain. A rough approximation of this depth is obtained by dividing the height of the water table above the drain as measured at the midpoint by 2. This gives, for the total thickness of the water-transmitting soil,

$$D = d + \frac{y_0}{2}$$

where d is the equivalent depth from the drain to the barrier as obtained from Hooghoudt's graphs.

The Application of the Formula in Irrigated Areas

The following discussion applies to the use of the formula by the Bureau of Reclamation.

A study of the ground-water hydrographs shows that the level of the water table fluctuates during the irrigation season. In general the water table is lowest at the beginning of the irrigation season and highest at the end. The highest points are reached immediately following an irrigation but the general trend is upwards throughout the season. During the nonirrigation season the water table recedes again and rises with the beginning of the next irrigation season. If the fall of the water table during the slack season does not equal the net rise during the irrigation season, the water table will have a progressive rise over the years. If the soil is to be drained properly this rise must be prevented.

Stated another way, the annual discharge from an irrigated area must equal or be greater than the annual recharge if the water table is to be maintained at reasonable levels. When the annual recharge is about equal to the annual discharge then the annual water-table fluctuations become reasonably constant from year to year. The Bureau of Reclamation defines this condition as "dynamic equilibrium."

In the use of the drain-spacing formula by the Bureau of Reclamation an attempt is made to take into account the varying regimen of the ground-water recharge and discharge. The goal of their analysis is to arrive at a drain spacing that results in a dynamic equilibrium with a specified water-table height under specific soil, irrigation, crop, and climatic characteristics of the area.

In order to perform the analysis it is necessary to know the consumptive-use requirements, the cropping practices, and the irrigation schedules and requirements.

The Bureau of Reclamation uses the concept of the specific yield. All the soil above the water table is assumed to be drained to the same moisture content. This is not actually true but is an approximation of the true moisture conditions. Regardless of objections to the specific-yield concept it is easy to visualize and apply to field situations. A relationship between the specific yield and the soil hydraulic conductivity has been developed by Talsma and

Haskew and extended by the Bureau of Reclamation for a number of different soils. The relationship must not be regarded as absolute one, for it is valid only for the soils for which the data has been collected.

There is a certain rationality about this relationship. The "specific yield" is a measure of the pore-size distribution in the soil. It is to be expected that a correlation exists between the pore-size distribution and the soil hydraulic conductivity. Although the relationship can never be better than an approximation, this approximation may well be within the deviation that one would expect from field measurement of the hydraulic conductivity.

The relationship used by the Bureau of Reclamation to estimate the specific yield from hydraulic conductivity is shown in Figure 10-10. In order to use this figure we must bear in mind that the value of the specific yield pertains only to the soil layer which is being drained. It is therefore necessary that the permeability measurements be made in this layer.

Calculations are started with the maximum height the water table will reach. This will occur at the end of the irrigation season after the last irrigation. It is necessary to calculate the average flow depth D for the maximum height that the water table will reach. First of all it is necessary to assume a value for L, the distance between drains. Then the values of k, t, and s which have been measured or calculated are used to follow the water-table fluctuations for the assumed spacing.

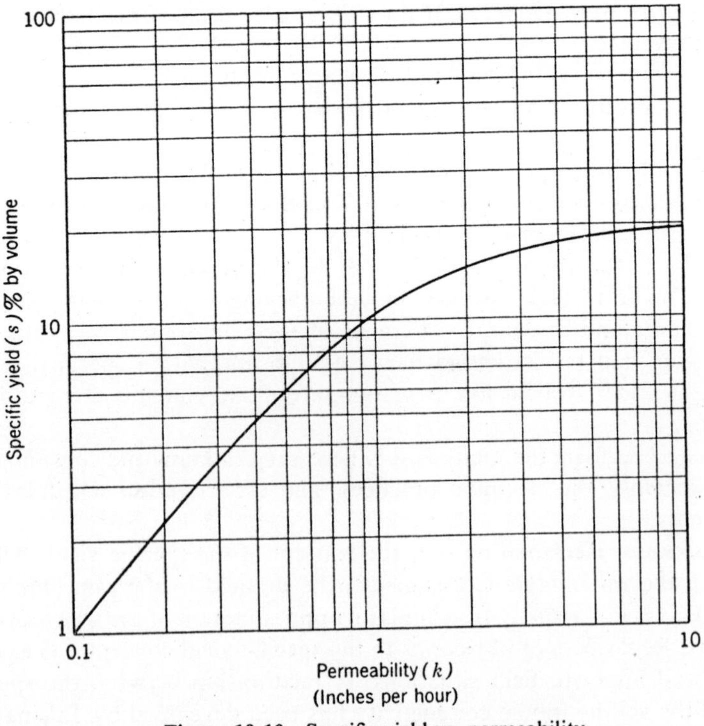

Figure 10-10 Specific yield vs. permeability.

It is possible to calculate the drop in the water table during the assumed time interval t by using the graphs. After each recharge the new position of the water table is calculated and then, after recharge, the drop is calculated. At the end of the season the water table should be in the same position or lower than at the start of the calculations. If the water table is higher, the drain spacing must be altered and the calculations repeated until the proper spacing is reached. Only two trials are normally required to estimate the proper spacing.

The Calculation of the Drain Discharge Using the Bureau of Reclamation Formula

The drain-discharge formula for a drain which is located above a barrier is

$$q = \frac{2\pi k y D}{L}$$

where q = drain discharge in cubic feet per linear foot of drain per day
k = the permeability in cubic feet per square feet per day and y, D, and L have the same notation as in the drain-spacing calculations.

The discharge formula for parallel drains on a barrier is

$$q = \frac{4k{y_0}^2}{L}$$

DRAINAGE OF ARTESIAN AREAS

Areas in which the drainage problem is caused by the upward seepage of water from a confined aquifer are, fortunately, not very common. In the United States these problems occur in the intermountain states and in localized areas throughout the western States. In the main, the problem of the drainage of artesian areas is limited to the irrigated regions located on alluvial soils having considerable stratification.

Several different formulas have been proposed. Some of the formulas take into account the spacing of drains for the situation of simultaneous upward seepage from the artesian aquifer as well as the downward seepage from rainfall or irrigation water. The formulas which result from these assumptions require a very close spacing of the drains. On the other hand if only upward seepage from the artesian aquifer is considered then the spacings are reasonable and are within the range of practicality.

A formula has been derived for the following set of conditions. It is assumed that a series of equally spaced tile lines are embedded in the soil at a constant depth below the soil surface. The soil is homogeneous and uniform with respect to the hydraulic conductivity and is underlain at a constant depth by an artesian layer. The water in the artesian layer is at a known pressure and steady-state conditions are assumed to exist.

The soil is assumed to extend an infinite distance above the drain lines. This assumption is contrary to fact and thus the water table heights predicted by the formula are slightly higher than the actual heights expected in the field. The formula is thus somewhat conservative for design purposes.

The water-table height at the midpoint between the drains is given by the expression

$$H = -\phi_a \bigg/ \ln \frac{\ln 2/(\cosh 4\pi h/S + 1)}{\dfrac{\cosh 2\pi r/S - 1}{\cosh 2\pi(r + 2h)/S - 1}} \tag{6}$$

where ϕ_a = hydraulic head in artesian aquifer
r = radius of drain
S = spacing of drains
h = distance from tile line to artesian layer
x, y = horizontal and vertical coordinates measured from the origin which coincides with the top center of a tile drain
ln = natural logarithm

Solutions to equation 6 have been computed and are plotted in the Figure 10-11.

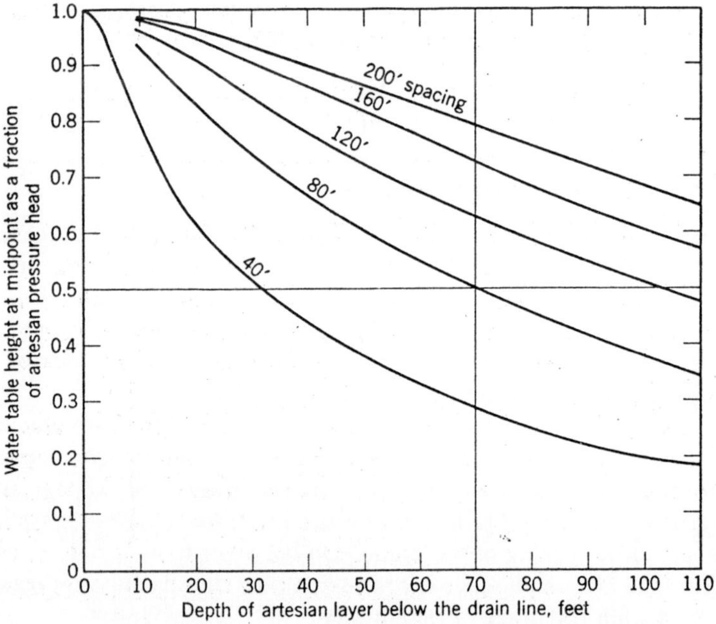

Figure 10-11 Chart for determining drain spacing in artesian areas. Radius of drain is 0.25 feet. All distances are with respect to the plane of the drain pipes.

The graph is used as follows: Determine the depth to the artesian layer and the artesian pressure head in the field. The depth of the tile line is frequently governed by factors which are not under the control of the designer, for example, the trenching machines which are available in the locality and the depth of the outlet. After deciding on the depth at which the tile is to be laid, the height of water table above the tile line necessary to give 3 feet (or whatever is deemed necessary) of unsaturated soil is calculated. The intersection of the abscissa and ordinate on the proper curve for the

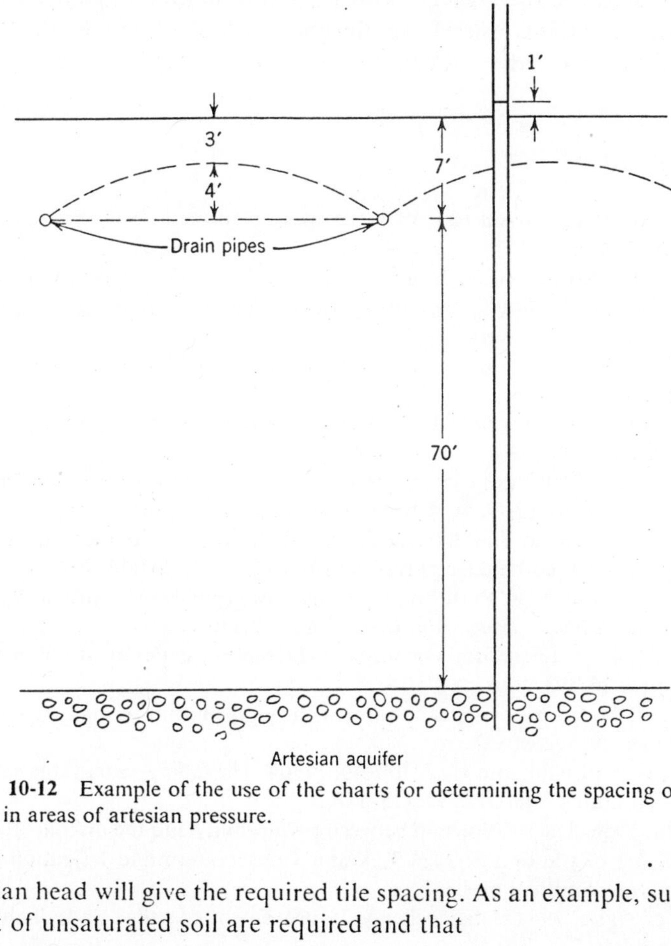

Figure 10-12 Example of the use of the charts for determining the spacing of subsurface drains in areas of artesian pressure.

artesian head will give the required tile spacing. As an example, suppose that 3 feet of unsaturated soil are required and that

1. The depth to the artesian layer below the tile line = 70 feet.
2. The artesian head with reference to the tile line = 8 feet.
3. The artesian head with reference to the soil surface = 1 foot.
4. The depth of the tile line = 7 feet.

The water table must not be permitted to rise more than 7.0 − 3.0 = 4.0 feet above the tile line. From an examination of the graphs we see that an 80-foot spacing of tile lines is correct.

The possibility of using pumped wells to drain artesian areas should be considered. There is also the possibility of using unpumped wells. If the water can be re-used, the economics favor the use of wells for the purpose of drainage. On the other hand if the artesian layer is a short distance below the soil surface the problem can be solved by placing a horizontal drain through it. In some instances drains as deep as seventeen feet have been used successfully to relieve the artesian pressure and to drain a large area. If the drain penetrates to the artesian layer, the graphs should not be used. Often a single drain will relieve the artesian pressure over an extensive area.

REFERENCES

Donnan, W. W. 1946. Model tests of a tile spacing formula. *Soil. Sci. Soc. Am. Proc.*, **2**:131–136.

Donnan, W. W., Aronovici, V. S. and H. F. Blaney. 1947. *Report on Drainage Investigations in Irrigated Areas of Imperial Valley, California.* Mimeographed report. U.S. Dept. of Agric.

Drainage-Vraagstukken, Voordracten 16 Mei 1952 van De Cultuurtechnische Vereniging.

Engelund, F. 1951. *Mathematical Discussion of Drainage Problems.* Danish Academy of Sciences No. 3.

Hooghoudt, S. B. 1940. Bijdragen tot de kennis van eenige natuurkundige grootheden van den grond, 7, Algemeene beschouwing van het probleem van de detail ontwatering en de infiltratie door middel van parallel loopende drains, greppels, slooten, en kanalen. *Versl. Landbouwk. Ond.*, **46**:515–707.

Isherwood, J. D., and A. F. Pillsbury. 1958. Shallow groundwater and tile drainage in the Oxnard Plain. *Trans. Am. Geo. Union*, **39**:1101–1110.

Kirkham, D. 1945. Artificial drainage of land: Streamline experiments. *Trans. Am. Geo. Union*, **26**:393–406.

Kirkham, D. 1958. Seepage of steady rainfall through soil into drains. *Trans. Am. Geo. Union*. **39**:892–908.

Ligon, J. T., D. Kirkham, and H. P. Johnson. 1964. The falling water table between open ditch drains. *Soil Sci.*, **97**:113–118.

Maasland, M. 1956. The relationship between permeability and the discharge, depth and spacing of tile drains. *Bull.* 1, Water Conservation and Irrigation Commission, New South Wales.

U.S. Dept. of Agric. 1954. Diagnosis and improvement of saline and alkali soils. Handbook 60.

Van Schilfgaarde, J., D. Kirkham, and R. K. Frevert. 1956. Physical and mathematical theories of tile and ditch drainage and their usefulness in design. *Res. Bull.* 436, Iowa Agric. Expt. Sta.

Van Someren, C. L. Undated. *Grafische Berekening Drainafstanden.* Cultuurtechnische Dienst, Ministerie van Landbouw, Visserijen Voedselvoorziening.

Visser, W. C. 1954. Tile drainage in the Netherlands. *Neth. J. Agr. Sci.*, **2**:69–87.

Wesseling, J. 1964. A comparison of the steady state drain spacing formulas of Hooghoudt and Kirkham in connection with design practice. *Tech. Bull.* 34, Inst. for Land and Water Management Research, Wageningen.

Wesseling, J. 1964. The effect of using continually submerged drains on drain spacings. *Tech. Bull.* 35. Inst. for Land and Water Management Research, Wageninged.

PROBLEMS

1. Use the Bureau of Reclamation graphs to compute the spacing required for the water table to drop from the soil surface to a depth of 1 foot in a 2-day period. The following information is available:

   The hydraulic conductivity is 1.8 inches per hour.
   Tile drains are to be placed 3.5 feet below the soil surface.
   The impermeable layer is 6.5 feet below the soil surface.

   What is the average flow out of a 200-acre field for the 2-day period?
2. Use Bouman's monograph to compute the drain spacing for the following situation:

   Steady rate of rainfall = 0.009 m/day
   Surface runoff = 0.001 m/day
   Deep seepage = 0.001 m/day
   Hydraulic conductivity = 0.001 m/day

   The drain pipes are placed at a depth of 1.2 m. The impermeable layer is at a depth of 2.5 m, and the water table should not be allowed to be closer than 70 cm from the soil surface.
3. Crops which can tolerate 6 mmhos conductivity in the drainage water are to be grown on an area of 3000 acres. The consumptive use of the crop is 38 inches of water. The winter rainfall is 6 inches. There is no other rain during the year. The conductivity of the irrigation water is 2 mmhos.
   (a) What is the leaching requirement?
   (b) What quantity of water must be drained from this acreage?
4. The E.C. of irrigation water is 1.3 mmho/cm. Assume a consumptive use of 0.35 in./day; a crop tolerance of 6 mmhos/cm; a soil hydraulic conductivity of 0.3 in./hour.

   The drains are to be placed at 8 feet and have a radius of 0.30 foot.

   The water table is not to be closer than 4.5 feet from the soil surface. The impermeable layer is 12 feet from the soil surface.
   (a) What is the drain spacing obtained by Hooghoudt's formula?
   (b) What will be the flow in cfs out of a 400-acre field?
   (c) If the outlet is on a grade of 0.001 what size of pipe is required?
   (d) If the water table rises to within 2 feet of the soil surface following an irrigation, how long will it take for it to drop to 4 feet below the soil surface (for the drain spacing calculated in *a*, using the Bureau of Reclamation charts)?

Chapter 11 SUBSURFACE DRAINS

A subsurface drain is one that is beneath the surface of the soil. It is buried out of sight and therefore has the important advantage of not interfering with the farming operations. The land can be farmed right over the drain and there is no loss of farming area due to the drainage system.

In general the initial cost of a subsurface drainage system is somewhat higher than the initial cost of an open ditch drainage system, but if one balances out the cost of the land removed from cultivation by the open drains, the costs of the two systems are often equal.

Figure 11-1 A mole plow used for draining land.

Figure 11-2 A well-laid curve in a tile line. Note the smoothness of the curve and the tightness of the joints. After Goodman. Photo courtesy Cornell University.

In general subsurface drains require less maintenance than do open drains. However, it is a fallacy to think that they do not require any maintenance. One of the disadvantages of a subsurface drainage system is the difficulty of finding out whether or not it is operating satisfactorily.

A wide variety of materials is used for subsurface drains. These materials include clay pipes in short sections, concrete pipe in various lengths, blankets of gravel laid in the soil, fibrous wood materials such as willow branches buried in the soil, covered stone drains, bituminous fiber materials, plastic pipe, and other materials which can be covered in the soil and which will remain intact for long periods of time. Mole drains are also included as subsurface drains. Mole drains are constructed by pulling a cylindrical metal object through the soil. The basic purpose of these drains is to collect the water that flows out of the soil and to carry this water into an outlet channel or conveyance structure. All of these subsurface drains are used for the same purpose regardless of the material from which they are constructed.

OUTLETS FOR SUBSURFACE DRAINS

There are two principal types of outlets used for farm drainage systems: gravity outlets and pump outlets. A gravity outlet is one in which the water flows out of the drainage system by gravity into a natural stream channel, an open ditch drain, a lake, a down well, or some other facility that disposes of the water. The topography of the land as well as the permeability of the soil dictates the choice of an outlet. The use of a gravity outlet presupposes that the elevation of the outlet is adequate to collect the water that flows through the drain.

Down Wells

Down wells, or down-draining wells are not commonly used. Their use is limited to areas of porous material such as fractured lava or porous limestone underlying the soil. An example of their use is in the North Side Minidoka Project in Idaho. The project is underlain by fractured lava having a high permeability. Drainage waters are collected in the shallow depressions or sumps adjacent to the wells and the water is allowed to fall into the well. Some provision must be made to remove the debris from the water before it is admitted into the well.

Before installing a down draining well it is desirable to check on the legal aspects of such an installation. There exists the possibility of polluting the ground water.

Discharge into an Open Drain

The usual outlets in a subsurface drainage system discharge directly into an open collector drain. The outlet pipe should consist of a solid pipe at least ten feet long. The pipe should extend over the water in the ditch so that the drainage water will fall directly into the water in the ditch, thereby preventing erosion. The pipe should be about 1 or 2 feet above the high-water level of

Figure 11-3 Drain sump for conveying surface water into a porous lava formation. Man in foreground is kneeling next to down-draining well. North Side Minidoka Project, Idaho.

Figure 11-4 Gravity outlet into main drainage ditch. Note how solid outlet pipe extends over water in ditch.

the ditch so that the end of the outlet will never be covered with water. It may be necessary to provide some protection for the bank at the point where the drain emerges. A wooden bulwark or a concrete bulwark will provide the required protection.

The open drains or collector drains may, in turn, carry the water to some natural channel permitting the water to return to the ocean or lake. The present trend in areas of high value land is to use solid pipes for the main drains. Undoubtedly this trend will continue to develop. The design of the collector drains then becomes important since they must be designed to carry all the drainage water.

Evaporation Sumps

In some areas there are no natural drainage ways or streams that can be used to carry away the drainage water. In other areas the salt content of the drainage waters precludes the discharge into existing streams that may be used for irrigation purposes. Under these circumstances the use of evaporation sumps should be investigated. These sumps are used in Australia along the Murray River. Consideration is given to the rate of evaporation in the

Figure 11-5 Drain sump discharging into an open drainage ditch. Man in background is standing on the sump. Imperial Valley, California.

area in designing the sump. If the sump is to cover a large area it should be borne in mind that the unit evaporation over a large area is smaller than over a small area. One of the largest examples of an evaporation sump is the Salton Sea in California. The entire drainage outflow from the Imperial Valley is discharged into the Sea. The sea has no outlet into the ocean. It is 30 miles long and 10 miles wide. The rate of evaporation from the Salton Sea is about equal to the rate of inflow plus the small amount of precipitation that falls on the area.

Pump Outlets

Pumps are used where the main drains are not sufficiently deep or where it is necessary to discharge the water from land below sea level into the sea. The size of the installation may vary from a small $\frac{1}{2}$-horsepower pump which discharges a few gallons a minute to a large installation which drains many thousands of acres and discharges thousands of second-feet of water.

On an individual farm, sump pumps are required because of the inadequate depth of the collector drain. There are several methods for collecting the water. The usual method is to construct a vertical sump. The individual drain lines discharge into the sump and a float-controlled pump periodically

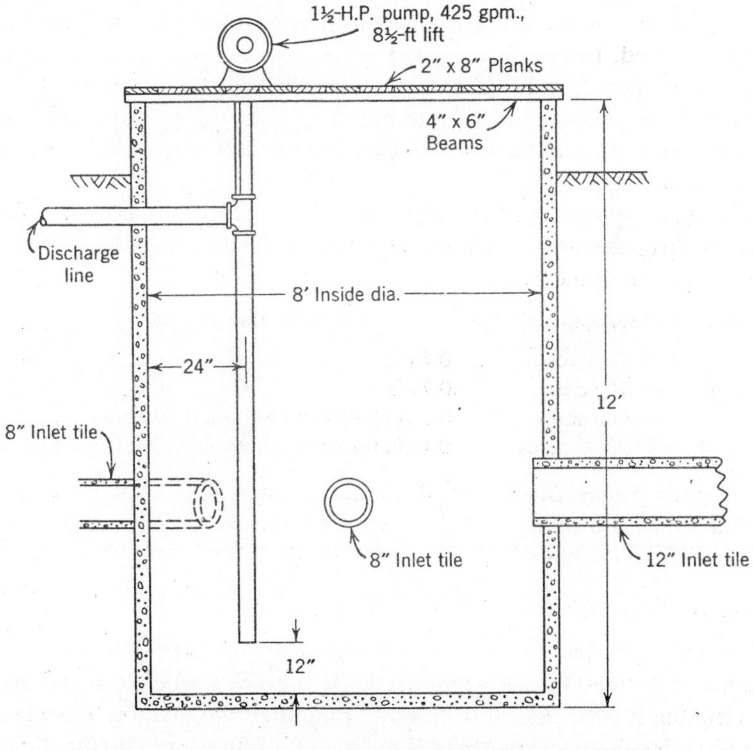

Figure 11-6 Sump-pump outlet, Imperial Valley, California.

empties the sump. Usually the water level in the sump is kept below the ends of the drain lines. However it is possible to allow the water level in the sump to fluctuate above and below the level of the drain lines. When this is done the storage capacity of the sump is increased by the amount of water stored in the drain lines.

Open drains on zero slope can serve as sumps also.

HYDRAULIC DESIGN OF SUBSURFACE DRAINS

The hydraulic characteristics of the materials used for drainage purposes must be known because they are used to carry the drainage water out of the field. The size of the drains must be adequate to carry the water at the design slope.

The Drainage Coefficient

The carrying capacity of drains has been investigated by Yarnell and Woodward in 1920, and an excellent chart has been prepared for the use in the design of a subsurface drainage system. The flow in the lines is based on the Manning formula with a roughness coefficient of 0.0108. This roughness coefficient is adequate for clay or concrete tile. It is probably not adequate for plastic pipe.

In order to use the chart it is necessary first to know the drainage coefficient. The drainage coefficient is the amount of water that must be removed in a 24-hour period. In humid areas the drainage coefficient usually varies from $\frac{1}{8}$ to 1 inch per day. It depends on the rainfall rate and the amount of surface drainage water that is admitted to the drainage system. The size of the watershed also affects the drainage coefficient as well as a number of other factors.

A set of design criteria for determining the amount of water to be drained from an irrigated area has been developed by measuring the outflow from existing drains in the area.

Acreage	*Required Drain Capacity*
0–40 acres	0.4 cfs
41–80 acres	0.7 cfs
81–900 acres	0.2 cfs for each additional 40-acre parcel
1,000–3,000 acres	0.1 cfs for each additional 40-acre parcel

The United States Bureau of Reclamation uses somewhat similar design criteria which are modified by the circumstances existing in the individual valley or drainage basin.

Slopes of Drain Lines

Subsurface drains can be laid on a variety of slopes. In some special instances they are laid with zero grade. However in order to make the lines self-cleaning it is necessary to have a grade that will give the flowing water enough velocity to carry the sediment out of the line. In the European

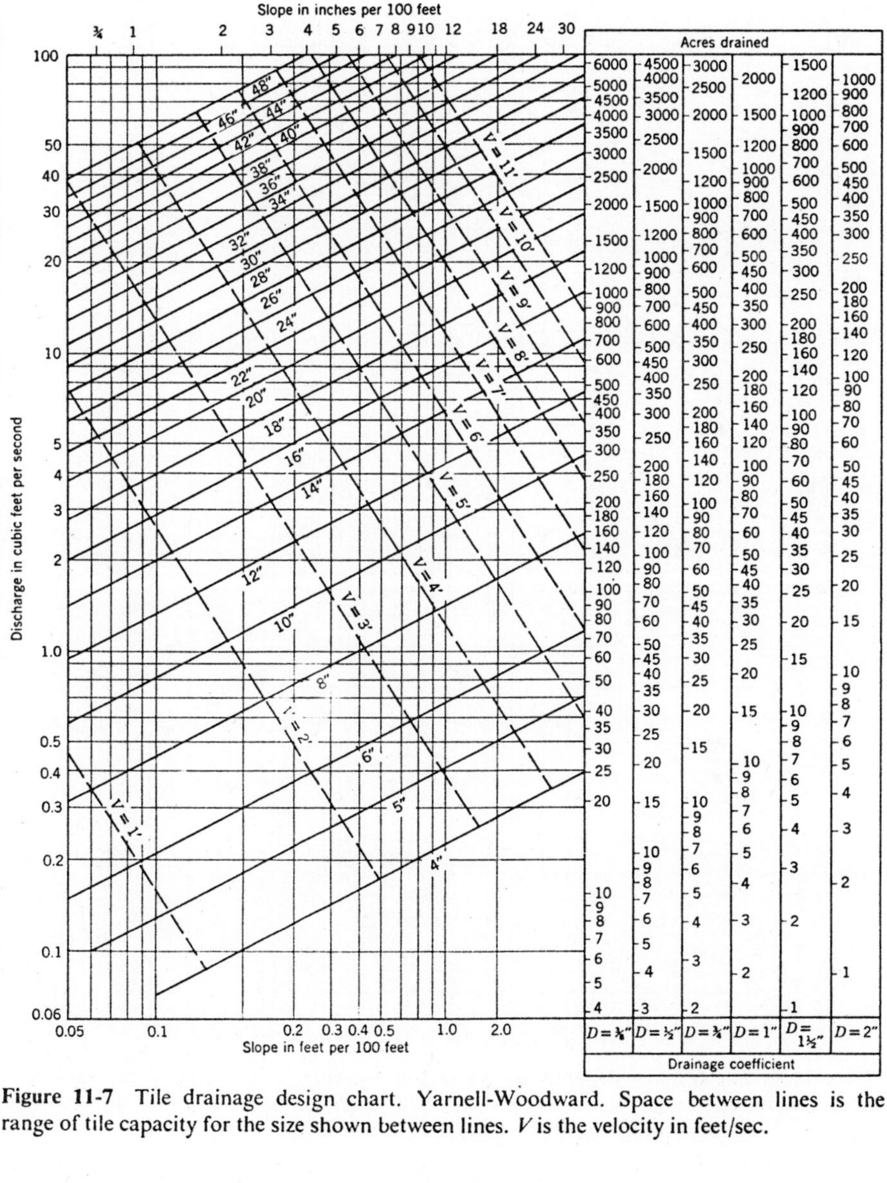

Figure 11-7 Tile drainage design chart. Yarnell-Woodward. Space between lines is the range of tile capacity for the size shown between lines. *V* is the velocity in feet/sec.

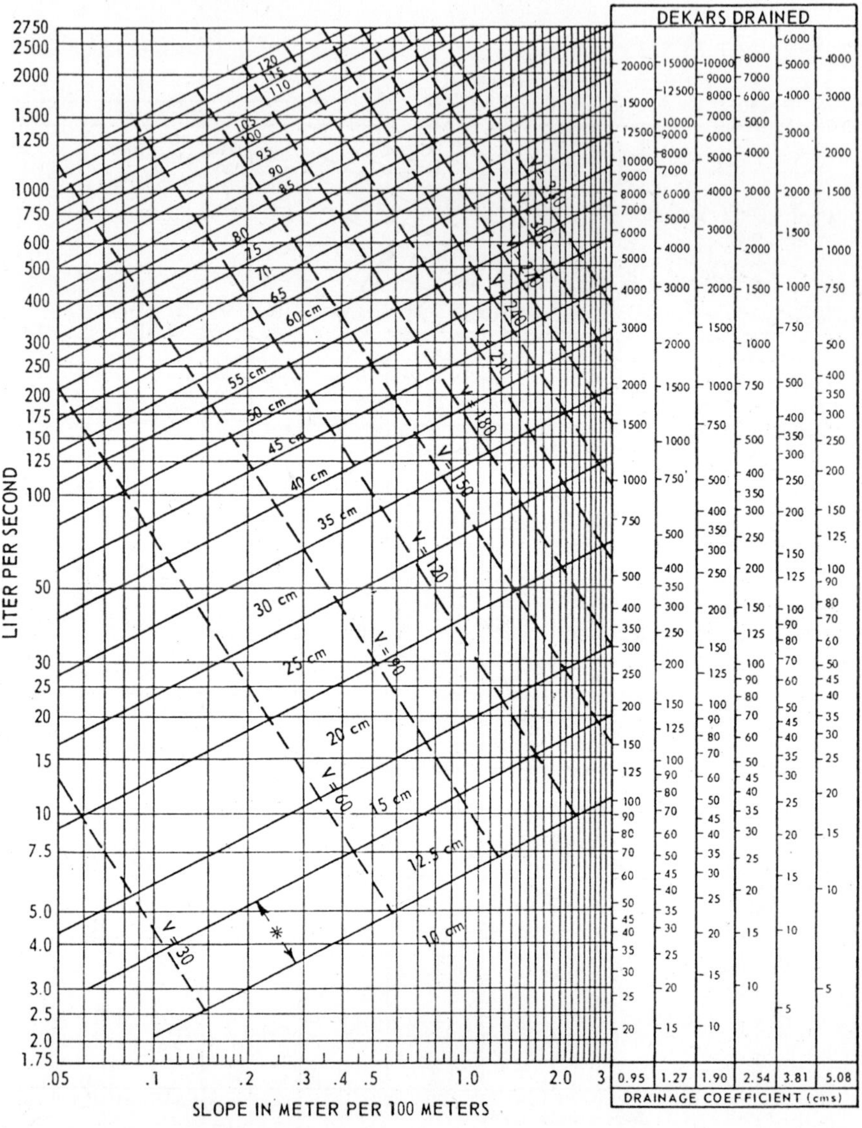

Figure 11-8 Tile drainage design chart. Space between lines is the range of tile capacity for the size shown between lines. From *Agricultural Engineers Yearbook*, 1959, transferred to metric system by Luthin-Tekinel.

Figure 11-9 Reducer on the end of a drain-discharge pipe. The purpose of the reducer is to reduce the flow into the main drain which consists of a closed pipe. The time required to lower the water table is increased. Red Cliffs, Australia.

literature a velocity of 20 to 25 cm/sec is considered to be adequate to carry the sediment out of the line. The American experiments indicate a higher velocity of 1 to 1.5 feet/sec (30–45 cm/sec). These velocities assume the drain line is running full. They may be taken from the Yarnell-Woodward chart.

Frevert et al. (1955) made an analysis of the problem of the minimum grade and recommend that it be 0.15% for 4-inch drain pipe to 0.05% for 12-inch lines or larger.

Slopes of 0.1% are difficult to maintain during construction by machinery. Studies of existing drain lines have shown many reversals of grade in the lines. Small reversals of grade are permissible, but if the reversal is greater than 10% of the inside diameter of the pipe, there will be reduction in the amount of water which can be transmitted by the pipe as well as increased danger of sedimentation in the pipe.

The Bureau of Reclamation specifications for the Riverton, Wyoming project states the following.

> The maximum allowable departure from grade shall not exceed 10% of the inside diameter of the drain pipe and in no case shall the departure exceed 0.1 foot. Where departures occur, the rate of return to established grade shall not exceed 2% of the pipe diameter per joint of pipe. The maximum allowable departure from alignment shall not exceed 20% of the inside of the drain pipe with a rate of return to the established line not to exceed 5% per joint of pipe.

ENVELOPE MATERIALS

A variety of materials have been placed around subsurface drains for the purpose of filtering fine sand and silt from the inflowing water. Gravel, coarse sand, organic materials such as corn cobs, safflower straw, and so forth, have been used for this purpose.

On the other hand many thousands of miles of subsurface drains have been installed without the use of any kind of envelope material and are working satisfactorily. In the middle western United States, in the states of Ohio, Iowa, and Illinois more than 20% of the farm land is drained by subsurface drains and little or no envelope material is used. The soil in these areas is stable and presents no problem. In irrigated areas the soil conditions can be quite unstable, especially at greater depths of installation and where the soil contains appreciable amounts of sodium on the base-exchange complex. Under these conditions it is essential to use some means of stabilizing the soil around the drain pipe to prevent its movement into the line. The greatest danger to the line occurs immediately after its installation. At this time the soil in the backfilled trench is in an unstable and fluid condition.

Criteria for the design of a gravel envelope or filter that will prevent the movement of fine sand and silt into the drain have been developed. Most of these criteria are based on the design of a gravel envelope for flow into a perforated well. The flow towards a well is much faster than the flow towards a subsurface drain. The design criteria are, therefore, quite conservative. The design is based on the ratio between certain size ranges in the soil and certain size ranges in the envelope material. The United States Bureau of Reclamation proposes the following criteria for the design of a gravel envelope.

$$\text{Uniform Material} \qquad \frac{\text{D50 filter}}{\text{D50 base}} = 5 \text{ to } 10$$

$$\text{Graded Material} \qquad \frac{\text{D50 filter}}{\text{D50 base}} = 12 \text{ to } 58$$

The soils in the Riverton, Wyoming project are very unstable and the Bureau of Reclamation recommends the following gravel for bedding purposes:

Retained in 5-inch screen	none
Retained in 4-inch screen	0%–20%
Retained in 3-inch screen	0%–30%
Retained in 2-inch screen	20%–50%
Retained in $\frac{3}{4}$-inch screen	20%–50%
Passing No. 4 screen	less than 5%

The gravel must be free of vegetable matter, bentonitic clays, and other deleterious substances.

In addition to the filtering properties of the sand or gravel mixture that is placed around the drain, there are several other factors which favor its use. The placement of gravel materials beneath the drain improves the bedding of the tile. Its bearing strength is increased since the weight of the tile line is distributed over a larger area of soil and the possibility of settling is reduced. It is only in unstable wet sands and silts that the danger of settling exists.

In some soils the flow of water into the drain lines is substantially increased by surrounding the drain with a thick layer of gravel or sand. When the construction of the system is done under high water-table conditions by machine the soil that is backfilled into the trench may be puddled. The soil's natural structure is destroyed and, in some instances, the permeability of the backfilled material is substantially reduced. The flow of water into the drain can be increased by placing a layer of gravel that fills the bottom of the trench.

There are a few isolated instances of the complete filling of the trench with gravel and sand. This is for the purpose of permitting the surface runoff to flow into the drain line. This practice is followed in the Napa Valley of California where the removal of the winter rainfall is a problem.

ENTRY OF WATER INTO SUBSURFACE DRAINS

In the case of subsurface pipe the water must enter through the cracks between successive pipes or through holes provided in the pipe for water entry. Some experimental porous pipe has been made but in general little or no water enters through the walls of the pipe.

Water enters mole drains over the entire permeable perimeter of the mole channels as well as the open wedge above the mole.

An analysis indicates that doubling the crack width will increase the flow into the drain, under ponded conditions, by about 10%.

In unstable soils it is usually recommended that the cracks be as small as possible. The irregularities at the ends of the segments of pipe are adequate to provide the necessary space for the entry of water into the drain line. In a few instances of pipe that has very smooth ends some difficulty has been experienced with the sealing of the cracks because of the lack of irregularities at the end of the pipe segments.

If envelope material is placed around the pipe, the pipe acts as though it were completely permeable. This is only true if the lateral permeability of the envelope is substantially greater than that of the soil. Where envelope materials are installed around the drain it is possible to use longer sections of pipe without any noticeable effect on the flow into the drain. Concrete pipe segments 30 inches long seem to give the same flow as segments which are 12 inches long, provided they are surrounded with a permeable envelope.

In humid regions there is often a recommendation for a crack width for pipe installed without an envelope. In stable soils the recommended crack width is generally $\frac{1}{8}$ to $\frac{1}{4}$ inch. In unstable soils the ends of the pipe segments must be butted together as closely as possible.

Effect of Drain Depth and Drain Diameter on Flow into a Drain Pipe

It has been shown (Kirkham, 1949) that as the drain depth increases so does the amount of water entering the drain. This conclusion is based on a study of the flow through soil having water ponded on its surface. Stated another way, the drain intake rate increases linearly with depth, the increase being somewhat greater at the smaller depth than at the larger depth. The diameter of the drain apparently has little effect on the flow rate. For example, at a 2-foot depth an increase in the drain size changes the flow rate as follows:

Drain Diameter	*Flow Rate*	*% Increase*
4 inch	0.287	—
6 inch	0.306	6.6
8 inch	0.317	10.5
12 inch	0.325	13.3

When the pipe is placed at a 12-foot depth the increase becomes

4 inch— —
6 inch— 8%
8 inch—14%
12 inch—23%

The above values are based on the assumption that the drains are running full with no back pressure. If the drains are running nearly empty the above increases will be larger, but they will still be relatively small. At a 4-foot depth a 400% increase of drain diameter (3 inches to 12 inches) increases the flow 20% when the drains are running full and 50% when running nearly empty.

Effect of Drain Spacing and Depth to Impervious Layer

Consider once again a soil with water ponded on its surface and underlain with drains equidistantly spaced through the soil. What effect does drain spacing have on the flow rate into the drain? Kirkham's solution for ponded water indicates that the flow rate into a drain is independent of the spacing of the drains as long as they are more than 20 feet apart. A change in the spacing of the drains does not affect the flow rate into a particular drain. Stated another way, the drain inflow rate is independent of drain spacings for all spacings ordinarily used. This means that, from a field having 50-foot spacings of drains twice as much water will be drained as from one having 100-foot spacings, and 4 times as much at 25-foot spacings as at 100-foot spacings. This statement is true regardless of how deep an impervious layer may lie, providing the depth of the impervious layer is the same for the whole field.

Although the drain inflow into an individual drain is independent of drain spacings for a normal spacing of drains, the inflow is not independent of the depth of the impervious layer unless the layer is several feet below the drain. If the impervious layer is more than 2 feet below the drain it has very little effect on the inflow rate into the drain. However, if the impervious layer is less than 2 feet from the drain the drain inflow rate is decreased accordingly. The fact that the drain inflow rate is decreased as the drain approaches the impervious layer does not necessarily mean that the drain should be placed above the impervious layer. In actual drainage practice it is desired to lower the water table as far as possible.

Effect of Placing Drain Pipe in a More Permeable Stratum

In laying out drain lines in the field the opportunity of placing the pipe in a more permeable stratum sometimes presents itself. It has long been realized that the flow into a drain line is increased by placing the line in a more permeable layer. For a soil with a more permeable stratum which is underlain by soil having the same permeability as a surface soil, it was found by Luthin (1953) that in order to double the flow into a tile line the ratio of the permeabilities, k_1/k_2, was changed from 1.00 to 0.18. In other words the drain-line stratum must be about $5\frac{1}{2}$ times as permeable as the rest of the soil to produce twice the flow that would occur in a homogeneous soil.

PERFORATIONS IN DRAIN PIPE

The effect of perforations on the flow into drain pipes has been studied by Kirkham and Schwab. As the number of holes is increased the relative increase in flow decreases, particularly if there are more than 20 holes per foot. Doubling the diameter of the perforations from $\frac{1}{4}$ to $\frac{1}{2}$ inch at the 4-foot depth increased the flow 68 and 46% for 4 and 10 holes respectively. The effect of perforations in reducing the flow is less at great depths than it is when closer to the surface.

The use of envelope materials around the perforated pipe means that there will be fewer perforations necessary in order to get the desired flow.

LOADS ON DRAIN PIPE

Pipe that is buried closer to the ground surface will be subject not only to loads due to the soil, but also to those imposed by equipment and animals on the ground surface. Pipe placed at greater depths will have only the load due to the soil above the pipe. Since the loads due to machinery are concentrated there is, in general, more danger of failure of shallow drain lines than of deep lines.

The manner in which the pipe is placed in the bottom of the trench is referred to as the bedding of the pipe. The resistance of the pipe to crushing can be increased greatly by the choice of the proper bedding procedure.

In order to determine the loads on drain pipe two formulas have been developed. One of the formulas, known as the projecting-conduit formula, applies to pipe placed in the bottom of a wide trench. The width of the trench is much greater than the outer diameter of the pipe. If the ditch is more

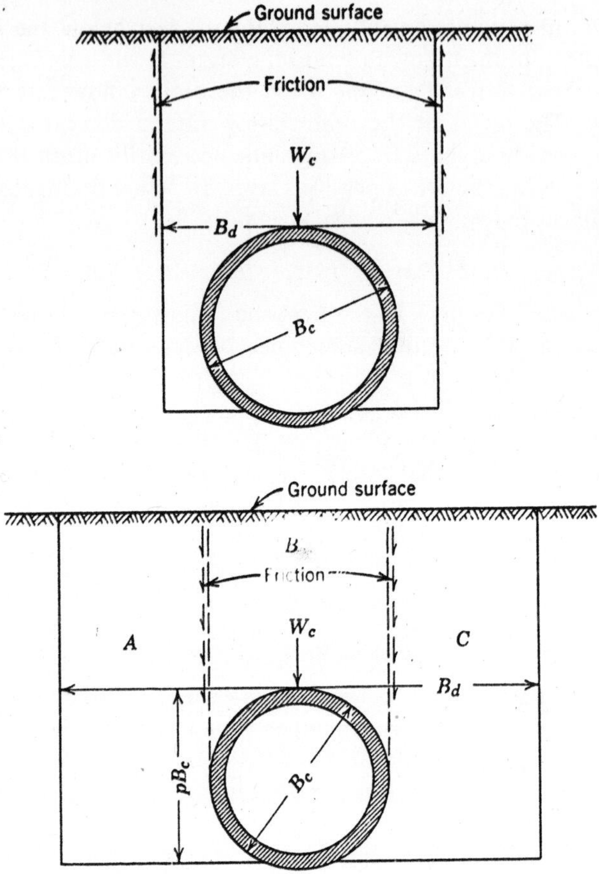

Figure 11-10 Top. Conduit in narrow ditch. Bottom. Conduit in wide ditch. After Van Schilfgaarde, Frevert, and Schlick.

than 2 or 3 times the diameter of the pipe, the projecting-conduit formula should be used.

For narrow trenches another formula known as the ditch-conduit load formula has been developed.

The Ditch-Conduit Load Formula (*Narrow Trench*)

The material which is used to cover the drain pipe is known as the fill material and it is assumed to have a lower bulk density than the soil adjacent to it. As the fill material settles into the trench there is friction between it and the walls of the trench. The load on the drain line due to the weight of the fill is reduced because of this friction with the walls of the trench. The formula for the loads on pipe in narrow trenches has been given by Marston (1930) as

$$W_c = C_d w B_d^2$$

where W_c = the total load on the pipe
C_d = the load coefficient
w = unit weight of fill material
B_d = width of the ditch at the top of the conduit

Projecting-Conduit Formula (Wide Trenches)

The projecting-conduit formula would apply to pipe laid in wide trenches as well as pipe laid in ditches made, for example, by a dragline. The load conditions in a wide trench are very different from those in a narrow trench. The friction between the walls of the trench and the fill material is unimportant. The material directly above the pipe will settle less than the material on

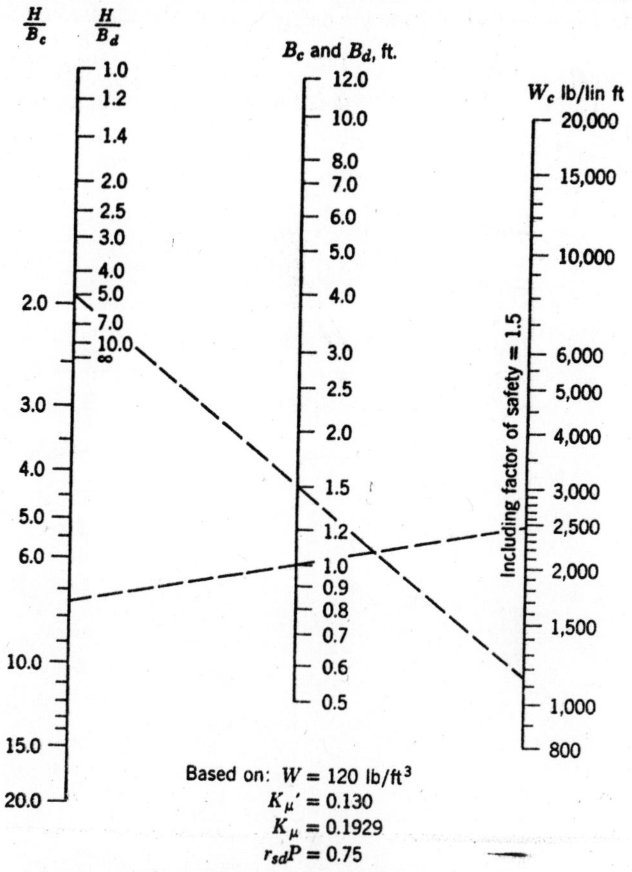

Figure 11-11 Nomograph for loads on tile installed in thoroughly wet clay.
H = depth of soil in ft above the top of the tile.
B_c = outside diameter of tile in feet.
B_d = width of trench above tile in feet.
W_c = load on tile in lb/lin ft.
(From Van Schilfgaarde, Frevert, and Schlick, Effect of Present Installation Practices on Draintile Loading, *Agric. Engr.*, 1951.)

either side of the pipe, and the load on the pipe is therefore increased owing to the settling of the material to the side of the pipe. The projecting conduit formula given by Marston (1930) is

$$W_c = C_c w B_c^2$$

where C_c = the load coefficient

B_c = the outside diameter of the conduit

The other symbols are the same as for the ditch-conduit load formula.

The Load Nomograph

The solution of the Marston equations is complex because it depends on soil properties which can be determined only with the use of specialized equipment. In order to facilitate the calculation of the loads, Van Schilfgaarde

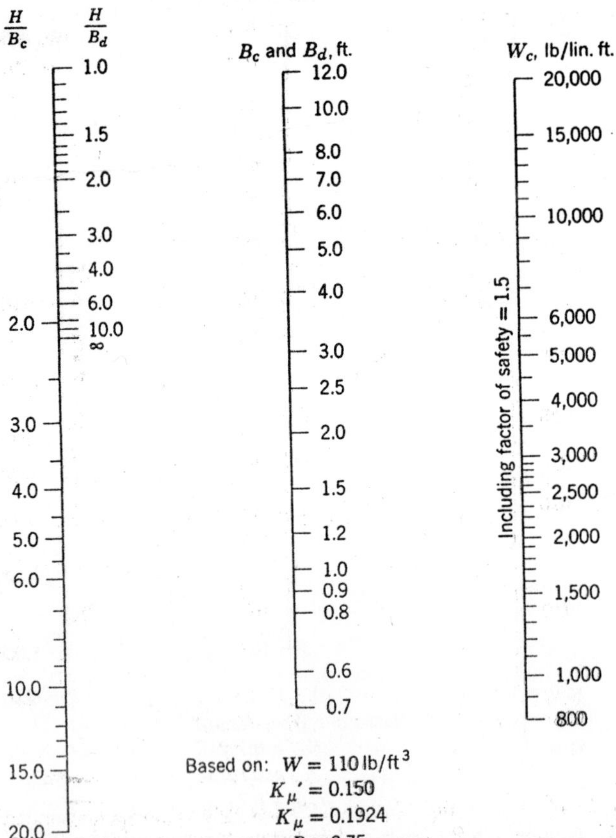

Figure 11-12 Nomograph for loads on tile installed in saturated sand. H = depth of soil in ft above the top of the tile; B_c = outside diameter of tile in feet; B_d = width of trench above tile in feet; W_c = load on tile in lb/lin ft. (From Van Schilfgaarde, Frevert, and Schlick, Effect of Present Installation Practices on Draintile Loading, *Agric. Engr.*, 1951.)

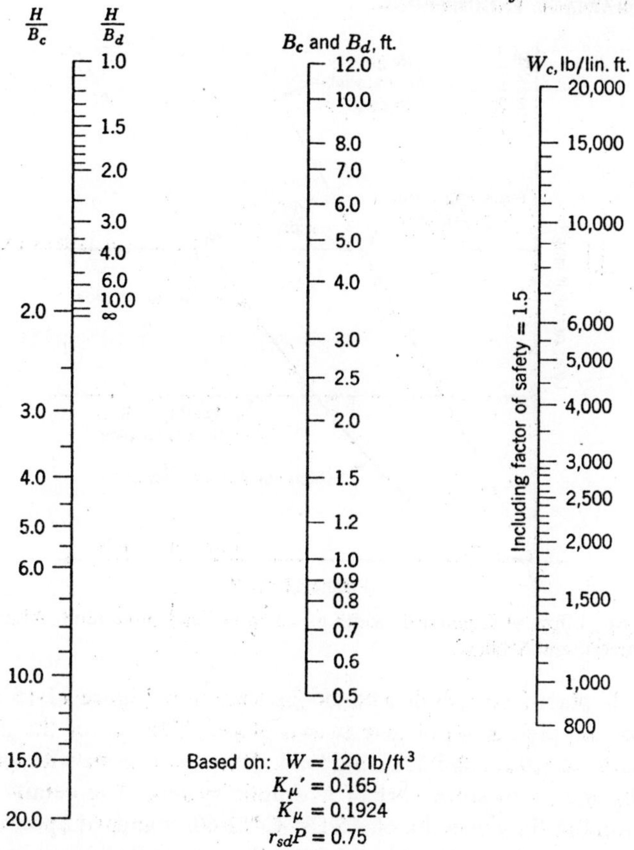

Figure 11-13 Nomograph for loads on tile installed in saturated top soil. H = depth of soil in feet above the top of the tile; B_c = outside diameter of tile in feet; B_d = width of trench above tile in feet; W_c = load on tile in lb/lin ft. (From Van Schilfgaarde, Frevert, and Schlick, Effect of Present Installation Practices on Draintile Loading, *Agric. Engr.*, 1951.)

et al. (1951) have prepared a series of nomographs. There are three nomographs, each one for a different soil. Although it is likely that the soil conditions in various localities will vary from those used in preparing the graphs, they can be used to obtain a satisfactory estimate of the loads to be encountered.

A plot of the two formulas to show the effect of an increase in ditch width on the load is shown in Figure 11-14. It is important to note that the load is calculated with each formula. The lowest value for the load is then used.

STRENGTH OF DRAIN PIPE

There are two methods in general use for testing the strength of rigid drain pipe. The first and most commonly used method is the three-edge bearing test. The equipment required for this test can be quite simple.

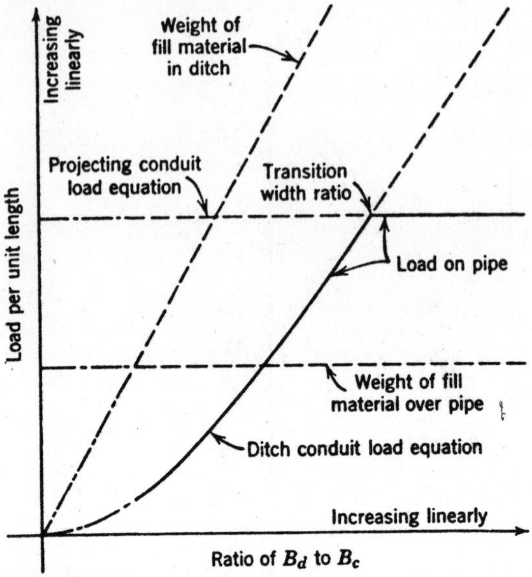

Figure 11-14 Effect of increase in width of ditch on load on conduit. After Van Schilfgaarde, Frevert, and Schlick.

The pipe is placed on wooden blocks as shown in Figure 11-15 so that the bottom of the pipe is supported at two places. The top of the pipe is then loaded with a continually increasing load. This can be done with a mechanical jack or by means of some sort of hydraulic system. The details of the test can be found in the appendix of ASTM C412-60, Standard Specifications for Concrete Drain Tile; ASTM C4-59T, Tentative Specifications for Clay Drain Tile, or in the Concrete Pipe Handbook published by the American Concrete Pipe Association (see references at the end of this chapter).

In the sand-bearing test the bottom part of the pipe is supported by a bed of sand. Although the sand-bearing test is less commonly used than the

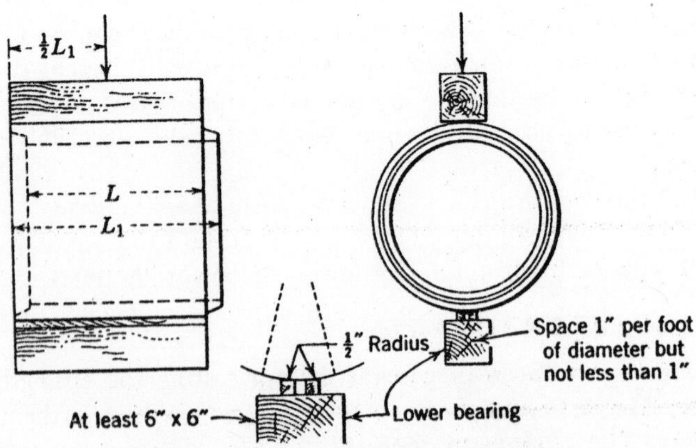

Figure 11-15 Three-edge bearing test.

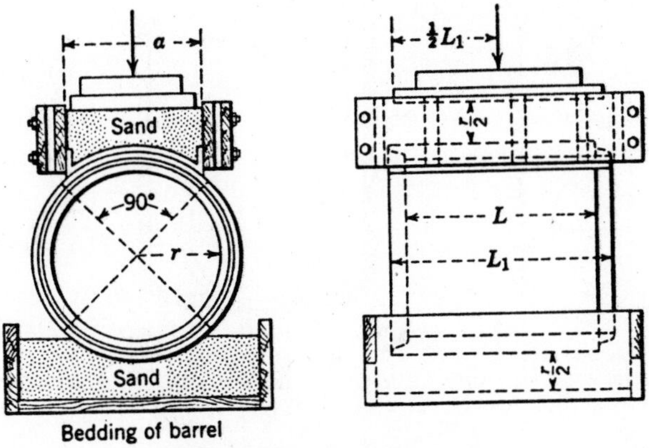

Figure 11-16 Sand-bearing test.

three-edge bearing test, the bedding conditions represented by the sand-bearing test more nearly approximate those found under normal installation procedures.

When subjected to the sand-bearing test a pipe will support more load than when tested with the three-edge bearing test. It has been determined that the crushing strength of pipe tested with the sand-bearing test will be 1.5 times the crushing strength of one tested with the three-edge bearing test.

BEDDING CONDITIONS AND THE LOAD FACTOR

The manner in which the pipe is bedded in the bottom of the trench has an influence on the load it will support. The load factor reflects this change in the amount of load the pipe will support. The crushing strength of the pipe as determined by the three-edge bearing test is multiplied by the load factor to determine the load the pipe will support under the particular bedding conditions.

A safety factor should be included in these calculations. This may be done by multiplying the calculated load on the pipe by 1.5.

Some of the types of bedding conditions are illustrated in Figure 11-17. Ordinary bedding is the one most commonly encountered in drainage installations. It approximates the conditions of the sand-bearing test, and the load factor is therefore 1.5. If the pipe is merely placed at the bottom of a level trench there is a 30% loss in the supporting strength of the pipe under these impermissible bedding conditions.

First-class bedding and concrete cradle bedding would not normally be encountered in farm drainage situations but would be desirable for the installation of drains in earth dams or along levees.

THE TESTING OF PLASTIC PIPE

Because of its recent introduction there are no well-developed criteria for plastic pipe such as exist for concrete and clay pipe. No specific recommendations can be given at this time, but it is interesting to examine some of the

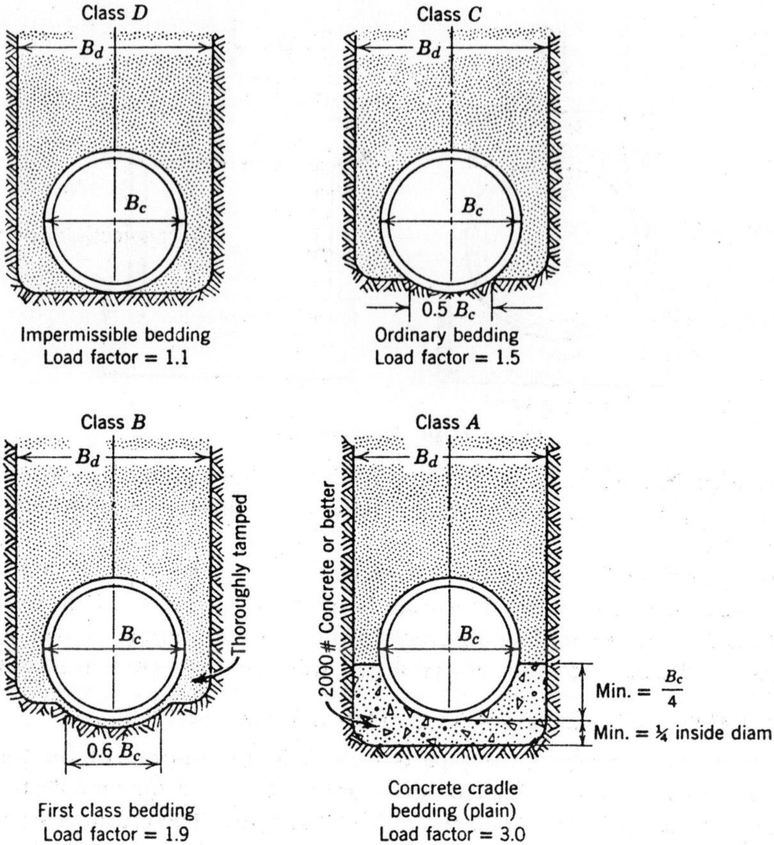

Figure 11-17 Various types of bedding conditions.

tentative requirements that have been established. In Holland the plastic pipe is laid in five meter lengths. It is coupled above ground and extended to the bottom of the trench. The bending radius must be less than 3 meters in order for this installation to be accomplished.

Plastic pipe does not crush as does rigid pipe. The deformation of the pipe is the property that must be measured. A tentative test in Holland is to place a 20-cm length of the pipe between two flat plates. A weight of 50 kilograms is applied for one minute. One minute after removing the weight the pipe must return to 90% of its original diameter. It is very important that these tests be carried out at 20°C.

At low temperatures plastic pipe becomes more rigid and therefore is subject to crushing. In order to test the properties of the pipe at low temperatures it is kept at 0°C for one hour. A 20-cm length is then placed in a 120° groove. A ball weighing $\frac{1}{4}$ kg and having a radius of $12\frac{1}{2}$ mm is dropped from a height of one m onto the pipe. Less than 5% of the pipes tested can be crushed in order to pass the test.

The reader must realize that the tests described above are only in the development stage and are not to be regarded as recommendations.

CHEMICAL DEPOSITS IN SUBSURFACE DRAINS

In a number of places in the world (California, Florida, Sweden, Norway, and Holland), red or black sludge-like deposits have occurred in subsurface drains. In some cases the drain is completely filled by the deposit and becomes inoperative. In other cases the deposit seals the joint or cracks between the pipes and prevents the entry of water into the drains.

It has been shown that these deposits are caused by the activities of bacteria which oxidize and precipitate reduced forms of iron and manganese in the drainage water. The iron precipitates are red in color and they have been observed in more instances than the black deposits which are caused by manganese.

Deposits have been observed in both clay and concrete pipe as well as in pipe made of bituminous materials. The deposits have also been observed in the envelope materials placed around the pipe. In some cases gravel envelopes have been completely clogged by the deposits. The clogging of glass fiber filters has also been observed.

In Florida, Spencer et al. report the complete clogging of some lines within a few months of installation. However, other lines installed in similar areas showed no precipitation even after seven years of operation. Soil analysis by these investigators indicated that the problem was most acute in the areas high in soluble iron in the reduced form.

The problem can be remedied by the use of 1*N* H_2SO_4 and 2% sodium bisulfite solution to dissolve the oxides of iron and manganese from inside the lines (Spencer and MacKenzie). In areas where these deposits are a problem it is desirable to install a breather at the end of each line to allow flushing of the line.

SPECIAL STRUCTURES FOR SUBSURFACE DRAINS

Manholes

Manholes are placed in the drain lines for the purpose of inspecting the operation of the line and trapping the silt that may be in the drainage water. In addition, a manhole provides for ready access to the line if repairs must be made to the line or if the line requires cleaning. The use of manholes is expensive. The cost may equal the cost of the rest of the line if many manholes are used. In general, manholes are only installed at the junction of the drain lines. In some instances where the roots of plants are a problem the manholes are spaced about 300 feet apart to permit periodic cleaning of the lines.

Manholes may be made of a variety of materials. The usual material is concrete but there are numerous instances where redwood has been used successfully. They should be sufficiently large in diameter to allow a man to descend to the drain line.

Manholes are objectionable to many farmers because of their interference with farming operations. In order to remove this objection a capped manhole is often installed. The top of the manhole is at least two feet beneath the ground surface so that it is well out of the zone of operation of farm implements. If inspection is to be accomplished it is necessary to dig out the top of the manhole. A capped manhole is not as convenient as an open manhole, but it is less expensive to construct and offers less obstruction to the farming operation.

Sedimentation Basin

A sedimentation basin can be incorporated into an inspection manhole. The usual practice is to have both the inlet and outlet pipe enter into the basin at level grade. If this is done there will be less neglect of the maintenance of the structure.

Blind Inlet

Blind inlets are used to allow surface drainage waters to percolate into the subsurface drainage system. Materials such as corn cobs, safflower straw, sand, gravel, and a variety of other materials can be used for blind inlets. In general the use of blind inlets has been unsatisfactory because they clog with fine soil particles and other sediment after a few years. The advantages of blind inlets are the initial low cost and the lack of interference with tillage operations. It is possible that the use of blind inlets will increase if better designs are found. In the Napa Valley of California blind inlets of sand and gravel extend all the way to the soil surface. These blind inlets have worked satisfactorily for a number of years.

Surface Inlet

A surface inlet admits surface water into the buried drain. In some cases surface inlets are installed for the purpose of flushing a drainline. A primary requirement for a surface drain is that provision be made for removing trash and suspended material from the drainage water. If this is not done there is the possibility of clogging the buried drain. If chemical deposits form in the drain line, the surface inlet can be used for chemical flushing of the line.

METHODS OF INSTALLING SUBSURFACE DRAINS

Hand Digging

The hand digging of drains is still done to a limited extent. The increased cost of labor has reduced the use of digging drains by hand. The use of hand labor for the digging of drains is limited to those areas where the drains are placed close to the surface. In the irrigated areas where the drains are placed at depths in excess of 5 feet all the digging is done by machinery.

Machine Installation

The main function of the machine is to dig the trench in which the drain is to be placed. There are two types of trenching machines in operation today.

One of these is the wheel type of excavator. The digging buckets are mounted on a circular wheel. As the wheel turns the soil is gathered into the buckets and dumped on a continuously moving belt that either dumps the soil at the side of the trench or is arranged to carry the soil to the rear of the caisson in which the drains are being laid. The soil then drops back into the trench to complete the backfilling of the trench in one operation.

Figure 11-18 Tile machine used for installing clay and concrete pipe. A man rides in the caisson and places the pipe in the bottom of the trench.

Another machine that is used for laying drain pipe is the ladder type of trencher. The buckets are mounted on a continuous belt that moves around straight sections of support. The ladder type of machine does not have as much difficulty from the soil caving in around the cutting section as does the wheel type of excavator. On the other hand the wheel type of machine is faster in its operation.

If the trench is more than 5 feet deep some protection must be provided for the man who works in the trench. The trench may be shored with removable planks. A less expensive method is to have a metal caisson on the machine. The caisson is to the rear of the digging buckets. All the drain pipe laying operations are accomplished in the caisson. A man sits in the caisson receiving the drain pipe from above and placing it on the floor of the caisson. A mechanical ram forces the pipe out of the rear opening in the caisson and keeps the individual sections of pipe from separating. Gravel may be fed through chutes to be placed around the drain pipe.

Figure 11-19 Ladder-type trenching machine used for laying drain pipe.

Figure 11-20 Drainage machine for installing plastic drain pipe in the Netherlands. The drain pipe is 40 or 50 mm in diameter and comes in lengths of 5 meters. The pipe lengths are coupled above ground.

Figure 11-21 Field laid out before the installation of concrete drain pipe. Note the main drain in the foreground and the laterals in the background. The piles of sand are used for envelope materials.

If such a machine is used to lay plastic pipe it is possible to feed the pipe from the top of the ground into the trench. The individual sections of pipe are coupled above ground and then fed to the bottom of the trench. In Holland the pipe length is 5 meters, and the pipe is sufficiently flexible so that it can be bent to go into the bottom of the trench.

Bituminous fiber pipe is installed in 8-foot lengths. It does not possess the flexible properties of plastic pipe and the caisson must be made long enough to accommodate the 8-foot lengths of pipe.

Any machine which can dig a ditch can be used for the installation of subsurface drains. Draglines are sometimes used as well as backhoes, scrapers, and so forth.

Recently there have been many attempts to make a machine that will pull the drain pipe into the soil in a manner similar to a mole plow. Difficulties have arisen in connection with maintaining grade but future developments may overcome these problems.

MAINTENANCE OF SUBSURFACE DRAINS

Too often the assumption is made that subsurface drains do not require any maintenance. It is true that in some areas subsurface drains have been operating for many years with little or no maintenance. However, in general the lack of adequate maintenance is a major cause of the failure of drainage systems.

Because subsurface drains are beneath the ground surface and are relatively inaccessible their maintenance presents some special problems. The inspection of the drainage system becomes difficult and is often neglected. There are several devices that permit periodic inspection of the drain lines. Most of

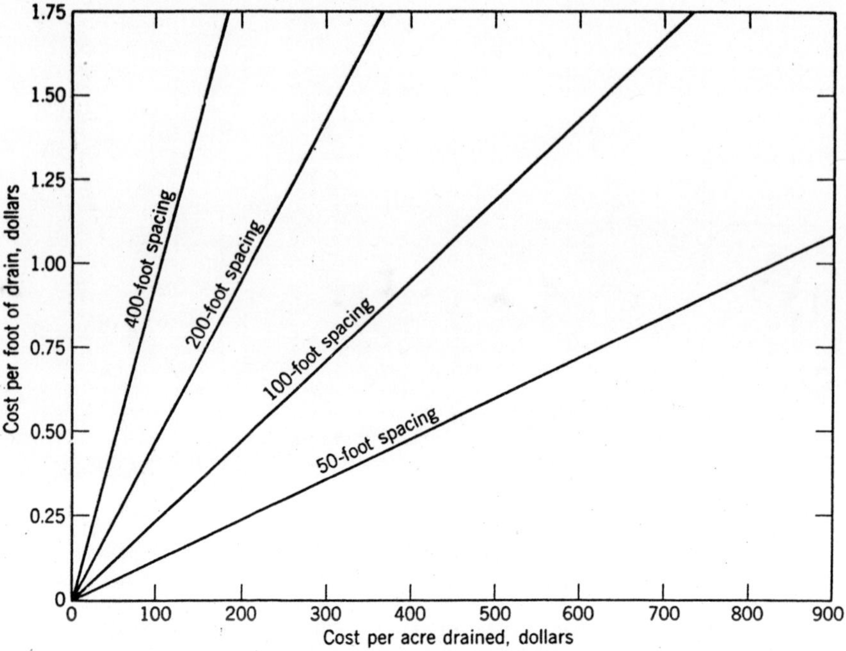

Figure 11-22 Graph for estimating the cost for installing drain pipe.

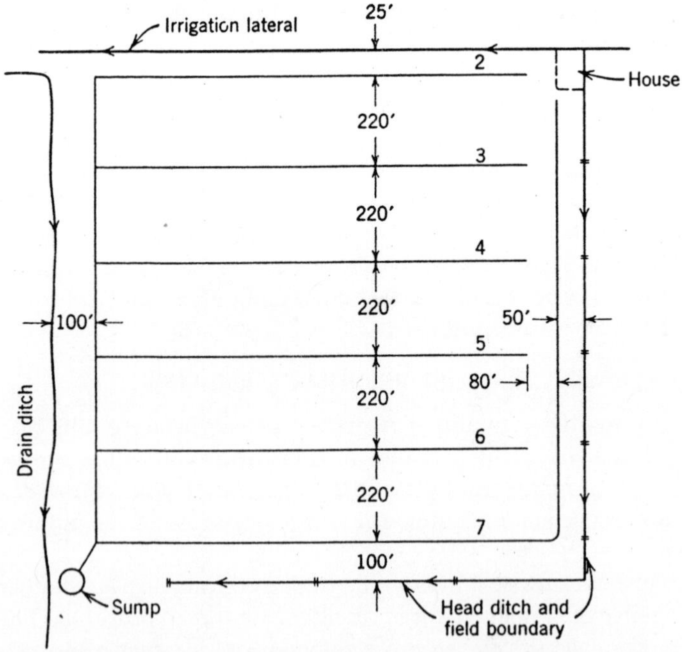

Figure 11-23 Typical drain installation. Imperial Valley, California.

these devices are above-ground installations that permit visual observation of the flow in the tile line. If possible, it is recommended that such an inspection device be incorporated in the drainage system.

In the event that a drain line is plugged by roots or by soil materials these can be removed with a mechanical rod made of flexible vanadium steel. Attachments can be fastened to the end of the rod to cut out the roots that may have plugged the line.

If the line is plugged with soil materials it is possible to use a hose that is specially designed for the purpose. A high-pressure stream of water is forced out of the end of the hose as it is shoved up the pipe from the outlet end.

REFERENCES

American Society for Testing Materials. 1959. Tentative specifications for clay drain tile. *ASTM*. C4-59T.

American Society for Testing Materials. 1960. Standard specification for concrete drain tile. *ASTM*. C412-60.

Averell, J. L. and P. C. McGrew. 1929. The reaction of swamp forest to drainage in Northern Minnesota. *Bull. Minnesota Drainage and Waters.*

Food and Agriculture Organization of the United Nations. Undated. Methods and machines for tile and other tube drainage. *Informal Working Bull.* 6.

Kirkham, Don. 1949. Flow of ponded water into drain tubes in soil overlying an impervious layer. *Trans. Am. Geophys. Union*, **30**:369–385.

Kirkham, Don. 1950. Potential flow into circumferential openings in drain tubes. *J. Appl. Phys.* **21**:655–660.

Marston, A. 1930. The theory of external loads on closed conduits in the light of the latest experiments. *Iowa Eng. Expt. Sta. Bull.* 96.

Roe, H. B. and Q. C. Ayres. 1954. *Engineering for Agricultural Drainage.* McGraw-Hill Book Co. New York.

Schwab, G. O. 1955. Plastic tubing for subsurface drainage. *Agr. Eng.*, **36**:86–89.

Schwab, G. O. and D. Kirkham. 1951. The effect of circular perforations on flow into subsurface drain tubes. Part II, Experiments and Results. *Agr. Eng.*, **32**:270–274.

Schwab, G. O., R. K. Frevert, and L. L. De Vries. 1956. Performance and operating costs of tile trenching machines. *Agr. Eng.*, **37**:469–472.

Spencer, W. F., R. Patrick and H. W. Ford. 1963. The occurrence and cause of iron oxide deposits in tile drains. *Soil Sci. Soc. Am. Proc.*, **27**:134–137.

Sutton, J. G. 1952. Maintaining drainage systems. *U.S. Dept. Agr. Farmer's Bull.* 2047.

van Schilfgaarde, J., R. K. Frevert, and W. J. Schlick. 1951. Effect of present installation practices on drain tile loading. *Agr. Eng.*, **32**:371–374, 378.

Visser, W. C. 1954. Tile drainage in The Netherlands. *Neth. J. Agr. Sci.*, **2**:69–87.

Weaver, M. M. 1964. *History of Tile Drainage.* M. M. Weaver, Waterloo, New York.

Yarnell, D. L. and S. M. Woodward. 1920. The flow of water in tile drains. *U.S. Dept. Agr. Bull.* 854.

PROBLEMS

1. A flat area is drained by parallel drain lines spaced 80 feet apart and 1000 feet long. If the drainage coefficient is $\frac{1}{2}$ inch what will be the flow at the outlet end of each drain?
2. If the field is 20 acres in size what will be the size of the main at the outlet of the field?
3. Design a gridiron drainage system for a 160-acre field with a slope of 0.015 on the laterals, and a slope of 0.010 on the main. Assume a drainage coefficient of $\frac{1}{4}$ inch and a drain spacing of about 100 feet.
4. Drains are spaced 150 feet apart in a field at a depth of 5 feet. Under ponded conditions the flow is in excess of the capacity of the main. What spacing is necessary to reduce the flow in the main by 20%?
5. A drain pipe is placed at the bottom of a trench that is 18 inches wide. Assume normal bedding conditions. The outer diameter of the drain pipe is 8 inches. If the soil is a loam what will be the load on the pipe?
6. What will the load be if the trench were 36 inches wide?
7. How many acres would a tile 8 inches in diameter on a gradient of 0.4 foot per 100-foot drain if a drainage coefficient (a) of $\frac{1}{2}$ inch (b) of $\frac{3}{8}$ inch (c) of 1 inch were used?
8. Laterals are spaced 100 feet apart in a given field. They have a fall of 0.3 foot per 100 feet, and are 1,500 feet long. Use a drainage coefficient of $\frac{1}{2}$ inch. What size tile would you use for laterals?
9. What slope is required to provide a velocity of 1.5 ft/sec at full flow in a 4-inch tile? In a 5-inch? In a 12-inch?
10. Select the grade for a 600-foot line tile outletting into an open ditch so that cuts do not exceed 4.2 feet. Design for an average depth of 4 feet. Elevation of the water surface in the ditch is 82.50, and hub elevations of successive 100-foot stations are 86.81, 87.29, 87.92, 88.15, 88.40, 88.61, and 88.79.
11. Determine the static load on 12-inch tile installed at a depth of 10 feet in a trench 24 inches wide, assuming that the maximum weight of saturated topsoil in the backfill is 110 p.c.f. Calculate the design load, using a safety factor of 1.5.

Chapter 12 OPEN DITCH DRAINS FOR CONTROLLING THE WATER TABLE

Open ditches are widely used for surface and subsurface drainage. They are used as individual field drains and for main drains. An important advantage of open ditches is their low initial cost. They are usually easy to construct, and a variety of machines is used for this purpose, as well as hand labor. An additional important advantage of open ditches is their ability to carry large quantities of water. They are the least expensive type of drain where large amounts of surface runoff caused by precipitation must be handled.

The commonest use of open ditches is as collectors or main drains for covered drainage systems, and as conveyance channels for surface runoff. They are also extensively used in forest swamp drainage of low-value land.

Open ditches are also used to control the water table. There are, however, some important disadvantages to their use in this regard.

The most important of the disadvantages is probably the difficulty they present for farming operations. Where the drains must be placed close together it is obvious that they will interfere with farming operations severely. An additional disadvantage is the maintenance. Maintenance of open ditches presents a constant problem. They must be maintained every year if they are to be kept operating. If they are not maintained they no longer act as drains. In many drainage districts where open ditches are the main type of drains, maintenance of the drains is not pursued vigorously enough and the ditches operate at a fraction of their effectiveness.

Another disadvantage of open ditches is the cost of land removed from cultivation. Figure 12-1 illustrates the value of the land removed from cultivation by open ditches. Still another aspect of open drains is the necessity for constructing bridges across them both for animals and vehicles. All these structures, and the open drains themselves, present a constant maintenance problem.

In addition to the above disadvantages of open drains, we can cite the difficulty of keeping them operating in unstable soil conditions. Open drains are subject to many different actions which cause them to fail.

In spite of these limitations, open ditches often present the only method by which drainage can be accomplished in an economical fashion. The designer must evaluate the advantages and disadvantages at each site in order to reach a proper decision.

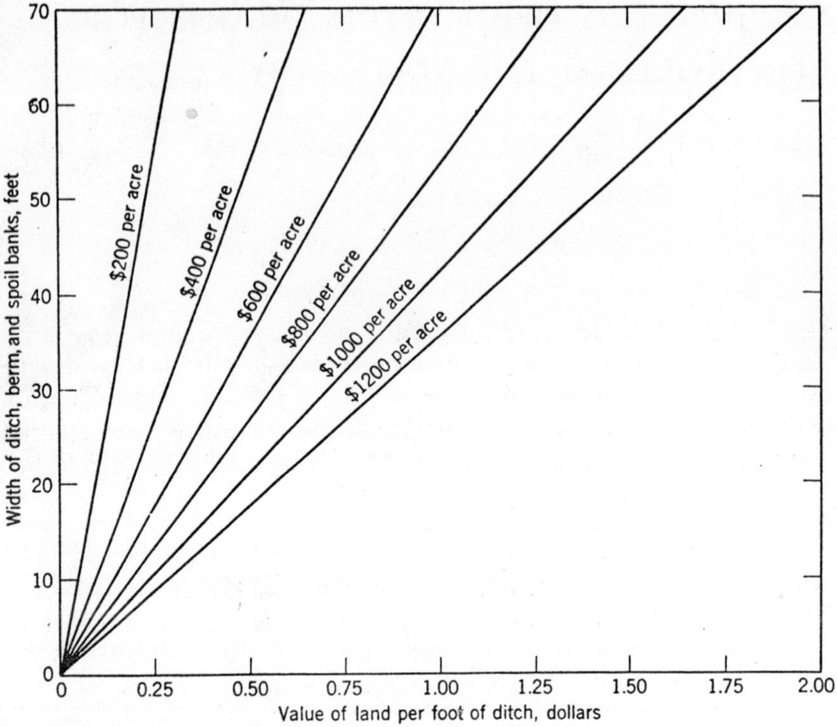

Figure 12-1 Cost of ditch drainage due to the cost of the land removed from cultivation.

DESIGN OF OPEN DITCHES

Channel Cross Section

The most efficient channel is the one that will have the maximum capacity for a given slope and cross-sectional area. The most efficient cross section is the one with the smallest wetted perimeter. This can more clearly be visualized by considering the Manning formula.

$$Q = av = \frac{1.486}{n} ar^{\frac{2}{3}} s^{\frac{1}{2}}$$

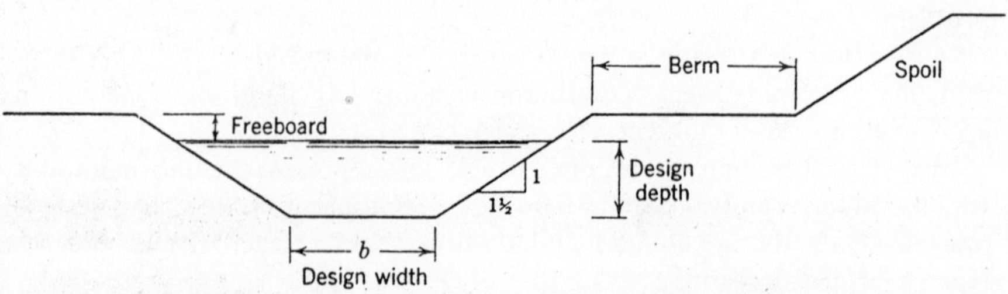

Figure 12-2 General features of a ditch for drainage purposes.

If we assume that a, n, and s remain constant for any ditch, then Q increases with r. Since $r = a/p$, r increases as p decreases. Therefore Q is maximum when p is minimum. Of all possible cross sections available to the ditch designer, the semicircle has the smallest wetted perimeter. Semicircles are sometimes used for concrete or brick channels but they are not used for earth channels.

Trapezoidal cross sections are used more than any other. Of all the trapezoidal sections the half hexagon has the smallest perimeter and has the most efficient cross section.

Side Slopes

A very important consideration in the construction of any ditch is the stability of the side slopes. The permanence of an open drain depends upon the stability of the side slopes of the drain. The characteristics of the soil have a great deal of influence upon this stability. Sandy materials which possess little cohesiveness are relatively unstable. Silt materials fall in the same category. When open ditches are dug in these materials the side slopes readily slough in because of the pressure of the water entering the sides of the ditches. This pressure of the water tends to lift the soil particles away from the sides of the ditch and cause them to fall to the bottom of the ditch.

Figure 12-3 Large open drain (17 feet deep) intercepting artesian flow from a layer about 15 feet below the ground surface. Moxee Valley, Washington.

Large sections of the bank may move into the ditch. An additional factor is the instability of the bottom of the ditch causing it to rise due to a differential pressure when the soil is removed from the ditch.

Various methods can be used for stabilizing the side slopes even though the soil material is relatively unstable. Tamping the sides of the ditch with the bucket of the drag line is one method by which this can be accomplished. Other methods include mechanical tamping of the sides of the ditch by special machines and rolling the sides of the ditch with weighted rollers.

Figure 12-4 Unstable ditch in sandy materials—Yakima Valley, Washington.

Side slopes vary from a 3-to-1 slope or greater in sandy soil, to a 1-to-1 slope, or almost vertical slopes, in clay and highly organic soils. For example, in the organic soils of the San Joaquin-Sacramento delta, the side slopes of the ditches are vertical. However, in certain other areas such as the Riverton area of Wyoming it has been virtually impossible to maintain open drains because of the unstable nature of the sandy materials through which they are being constructed. Artesian pressures also make it difficult to maintain open drains.

Scouring and Deposition

Scouring and deposition in a ditch depend upon the soil material through which it is constructed and upon the slope or grade of the ditch. The slope or grade of the ditch must be adjusted to suit the particular soil conditions through which it is being constructed. The best procedure in the construction of an open ditch is to observe similar open ditches which have been constructed in the same area and see how they are operating for particular slope and soil conditions. However, if such ditches are not available it is advisable to rely upon the guides prepared by governmental agencies such as the Soil

Figure 12-5 Structure for controlling erosion in soft limestone formation and for maintaining a high water table. Southwest of South Australia.

Conservation Service. These guides give the maximum permissible grades for ditches constructed through various types of soils. Scouring and deposition are increased by changes in grade of the ditch. A flattening of the slope of the ditch will cause deposition, whereas an increase in the slope of the ditch may cause accelerated scouring or erosion.

Disposition of Spoil

Spoil is the earth removed in the construction of the ditch. Usually the spoil is piled some distance away from the side of the ditch. The area between the spoil bank and the side of the ditch is known as the berm. The width of the berm will depend on the depth of cut, the stability of the excavated section and the type of maintenance equipment to be used. The stability of a ditch bank is reduced if the spoil is placed adjacent to the side of the ditch.

The additional weight of the spoil increases the hazard to the stability of the ditch.

Spoil banks are generally unsightly, and often reduce the farming area unnecessarily. It may be possible to spread the spoil over the field. Consideration should be given to the fertility and salinity of the spoil material.

TABLE 12-1 Safe Velocities for Various Soils

Soil	*Maximum Velocity, fps*
Sandy and sandy loam (noncolloidal)	2.5
Silt loam	3.0
Sandy clay loam	3.5
Clay loam	4.0
Heavy clay, fine gravel, and graded, loam to cobbles	5.0
Graded, silt to cobbles (colloidal)	5.5
Shale, hardpan and coarse gravel	6.0

Taken from the Iowa Drainage Guide

If the salinity levels are high and the fertility low there can be a reduction in the overall yield of the field unless corrective measures are taken such as the addition of fertilizer or the leaching of the spoil material.

MAINTENANCE OF OPEN DRAINS

An open drain must be maintained properly if it is to do the job for which it was intended. A neglected drainage ditch will soon fill with weeds and sediment and its function as a drain will be severely reduced. Yearly maintenance is necessary to keep the drains in operating condition. The best type of maintenance is preventive maintenance—correction of the problem before it becomes serious.

Control of Sediment in an Open Drain

In humid regions most of the sediment that comes into the drain is due to erosion of the surrounding land. Erosion control measures can help a great deal in reducing the seriousness of the problem. On the other hand there is often scouring and deposition occurring within the ditch itself. This is particularly true where it is necessary to change the slope of the drain. The steeper reaches will erode and the material will be deposited on the flatter slopes where the water is moving slowly. In many cases it is not possible to prevent this accumulation of silt in the drain, in which case the drain must be periodically cleansed of silt. The use of settling basins or low check dams is recommended in conjunction with the open drain. The silt is concentrated in the settling basin or behind the check dam and can be removed easily.

Under certain special circumstances it is possible to use the silt to build up the level of lower lying land. The possibility of this will depend on the topography of the farm. By spreading the waters out over the land in a thin sheet the silt will be more or less uniformly deposited over the surface.

Bank-Erosion Control

In flat areas the erosion of the bank is only a slight problem. Velocities of flow are not large and it is usually possible to lay out the ditches along straight lines.

In regions having appreciable land slope the control of bank erosion becomes an important consideration. Rock or brush revetments should be used to control the bank erosion. These can be quite expensive to construct. Rock jetties can be placed in the ditch to deflect the water away from the bank which is being cut. Care must be exercised in the placement of jetties since it is possible to deflect the water too much, thereby causing erosion of the opposite bank.

Control of Side-Slope Damage

It is essential to provide the proper side slope during the construction of the drain. The earlier practice of assuming that the ditch can be cut with steep side slopes originally should not be followed. The theory was that the sides of the ditch would assume their natural slope and the bottom would partially fill with sediment. It has been adequately shown that this practice is faulty. The sides of the ditch cave in and leave a steep side slope. The ditch is partially filled with soil and no longer operates at the design capacity. It is much better to provide for a conservative side slope during the construction of the ditch. In some cases where large amounts of flood water must be carried off of sloping land it is possible to use a very gentle side slope. The side slopes can then be planted with grass to prevent erosion.

Weed Control

The problem of weed control is especially important. The growth of weeds in the ditch can seriously reduce the capacity of the ditch to carry water.

Weeds can be removed by chemical means or by mechanical methods. The recent development of weed killers has greatly facilitated the control of aquatic weeds. The use of high temperature burning has also reduced the labor involved in the control of weeds. Extensive literature is available from the manufacturers on the use of chemicals and burners for the control of weeds.

Some weeds such as tules can be controlled by mechanical procedures. A chain is dragged by two tractors; one on either side of the ditch. This dragging is done when the tule plants are just beginning to grow. The young plants are pulled out of the ground by the chain and the control is effective if performed every year.

METHODS OF CONSTRUCTION

Hand Ditching

Labor has become so scarce in most countries that the use of hand labor for the digging of ditches has almost disappeared. There are undoubtedly countries where the use of hand labor is still justified on the basis of cost. In general, hand-dug ditches should not have a bottom width of more than 3 feet or a depth of more than about 4 feet. In most instances it is more economical to use machines for digging ditches.

Drain Plows

For the drainage of low-value land, and especially for the drainage of forest land the use of plows is recommended. Special large-size plows have been developed in Finland for this purpose. The plows are almost 2 meters high and can dig a trench about 1 meter deep. Power requirements for the

Figure 12-6 A large plow constructing a ditch for forest drainage. Finland.

operation of the plow are high, and in most instances it is necessary to winch the plow through the soil. Large-size tractors such as the D-8 (Caterpillar) are normally used to supply the power. Under Finnish conditions the ditches are dug in soft materials such as peat swamps.

The ditch that is made by the plow is somewhat rough but the value of the crop does not justify more expensive methods. The Finnish plow is designed to leave a berm on either side of the ditch.

Other types of ditching equipment including scrapers can be used in manners similar to those employed with the drainage plow.

Crane Shovels

A crane shovel consists of a lower travel assembly, which can be either a truck assembly or a crawler, and an upper revolving frame mounted on the lower travel assembly. On top of the upper revolving frame is mounted one of six basic units which are called front end attachments. The front end attachment is more or less permanently mounted on the revolving frame and the final piece of equipment takes its name from the front end attachment. The six front end attachments commonly encountered are:

1. Shovel front
2. Crane
3. Clamshell
4. Dragline
5. Backhoe
6. Pile driver

In drainage work the clamshell, dragline, and backhoe are the only front end attachments used to any extent.

CLAMSHELL. The clamshell bucket consists of two halves, or shells, hinged at the top so that the bucket can be opened. The hinging permits the shells to be drawn together to form a bowl-like arrangement. The bucket is dropped on the material to be excavated with the shells open. The dropping of the bucket allows it to dig into the material to be excavated. The closing line brings the two halves of shell together. As the shell closes, the cutting edges dig into the material filling the bucket. The material to be dug must be relatively soft or loose since the cutting results from the weight of the bucket.

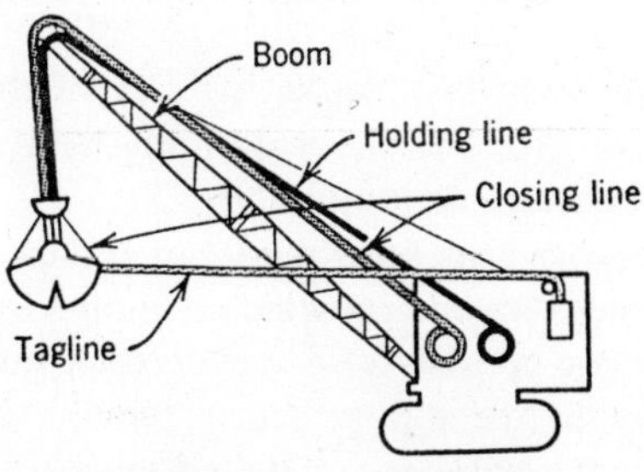

Figure 12-7 Clamshell.

Clamshells may be used to dig straight down, and are very useful for digging sumps or for digging in trenches that are sheathed or have crisscross bracing. They are excellent for underwater work.

DRAGLINES. The dragline gets its name from the dragging action used to load the bucket. The bucket is a scoop that is tossed away from the machine into the material to be excavated. It is then filled by being dragged towards the machine. Draglines are best used for the excavation of loose materials that are below the grade of the machine. They can also be used in underwater excavations. Draglines are probably the commonest machine used to dig drainage ditches. They are especially suited for digging beneath the water table. The machine remains on dry ground while digging. If the ditching is done in swampy areas, large rigid timber mats are fabricated and fitted with lifting slings. The operator can lift one mat and place it ahead of the machine when movement is necessary.

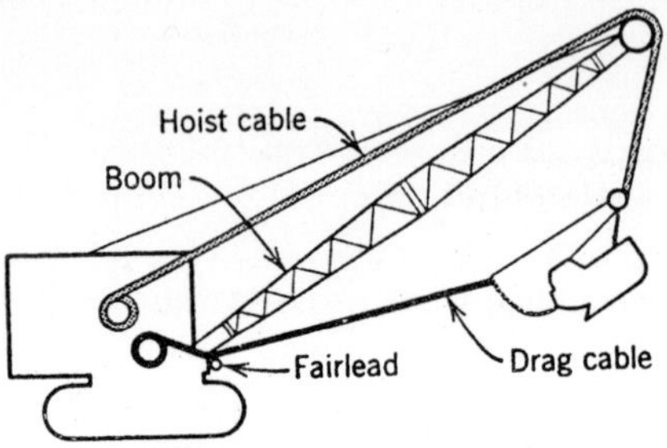

Figure 12-8 Dragline.

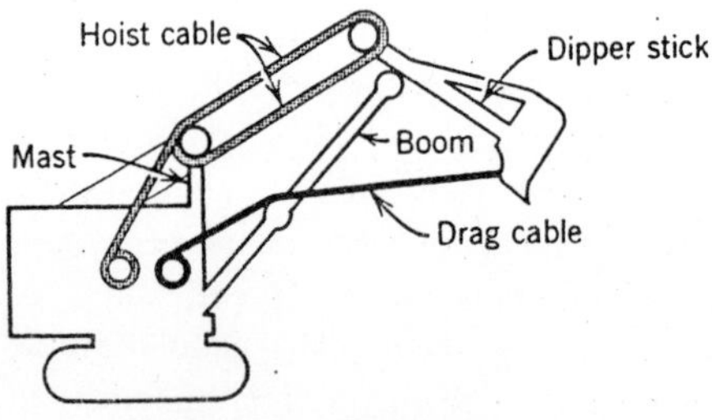

Figure 12-9 Backhoe.

The dragline can deposit the spoil on either side of the ditch or on both sides. It is the best machine for casting or dumping material as far from the machine as possible, because of its long boom.

BACKHOE. The backhoe is also known as the trench hoe, hoe, or drag shovel, because the dipper is on an arm which is pulled towards the machine when digging. The backhoe consists of a dipper stick pivoted at the end of the boom. The boom, with the dipper arm extended, is lowered into the excavation. The dipper is then dragged or pulled towards the machine by the dragline. When the dipper is filled the entire assembly is raised by the hoist line. The load is deposited by extending the dipper stick.

The backhoe incorporates some of the characteristics of the shovel and the dragline. Its primary use is for digging below the ground level. It will dig harder material than the clamshell or dragline because the weight of the boom itself is used to force the dipper into the material. It is possible to dig more accurately with the backhoe than with the dragline. It is an excellent machine to use when precision is needed in digging.

WORK OUTPUT FOR CRANE SHOVELS. The formula which is used to estimate the work output for crane shovels is as follows:

$$\text{output} = \frac{3{,}600 \times Q \times f \times E \times k}{C_m}$$

where 3,600 is the number of seconds in an hour.

Q is the rated dipper, clamshell, or dragline-bucket capacity in cubic yards.

f is the soil-conversion factor necessary to convert the bucket capacity into actual volume of earth removed.

E is the shovel-efficiency factor that takes into account the fact that a full 60-minute hour work time is rarely accomplished. For average operators the efficiency factor is about 50%.

C_m is the total cycle time in seconds. The working cycle is made up of digging, swing to dump, dumping, swing to dig.

k is the bucket or dipper-efficiency factor. It takes into account the variation of the bucket capacity when used in different soil conditions. For easy digging in loose soft materials the efficiency factor is from 95 to 100% for both the dragline and the shovel dipper. Medium digging in hard material that does not require blasting will result in efficiency factors of 85 to 90% for shovel dippers to 80 or 90% for draglines. Most drainage work would be in soft loose materials.

Some cycle times for draglines in the sort of material encountered in drainage work are given in Table 12-2

TABLE 12-2 Cycle Times for Digging in Different Materials

		Cycle Time (sec)		
Equipment	*Capacity* (cu yd)	Easy digging (light moist clay or loam)	Medium digging (loam)	Hard digging (hard tough clay)
Dragline (90° swing)	$\frac{3}{8}$	20	24	30
	$\frac{3}{4}$	22	26	32
	$1\frac{1}{4}$	24	28	35
	2	28	33	40
	4	32	36	45

For each increase of 10° in swing add 2 sec to the cycle time. For each decrease of 10° in swing subtract 2 sec from the cycle time. The conversion factors for certain soils are as follows:

sand	1.11
loam	1.25
clay	1.43

These conversion factors apply to material which is transformed into a loose condition by digging. In other words 1 cu yd of sand will occupy 1.11 cu yd in the bucket.

Stage Construction

Any of the digging procedures just outlined can be carried out in stages. Stage construction is used in unstable soils. If the ditch were dug initially to the finished depth, the sidewalls would collapse and the ditch would fill with soil. The collapse of the sidewalls would be due to the high water-table conditions that result in excessive seepage pressures on the wall of the ditch. Under these circumstances it is necessary to dig a shallow ditch first. The water table will then be lowered by this shallow ditch and the seepage pressures will be reduced. On the next pass of the machine the ditch is deepened once more. The number of times this process must be repeated can be determined only by experience.

During the digging of the ditch it may be possible to increase the stability of the ditch bank by tamping the soil with the bucket or with special rollers.

THE RELATIVE EFFECTIVENESS OF OPEN VERSUS SUBSURFACE DRAINS

The question of the relative effectiveness of open drains versus subsurface drains inevitably arises in connection with the selection of the best method of drainage. Theoretical considerations, such as Kirkham's solution for the ponded-water case, would indicate very little difference in the effectiveness of the two types of drains. It might be expected that an open drain would lower the water table more rapidly than would a subsurface drain because of the greater size of the drain. Kirkham has shown that an increase in the diameter of the drain will create an increase in the rate of flow into the drain.

For interception purposes a subsurface drain may be at a disadvantage compared to an open drain. On sloping land it is possible to have capillary flow above a subsurface drain. The drain will not intercept this flow. On the other hand an open drain will intercept all of the flow occurring in the region above the invert of the drain. All of this assumes that the open drain is maintained in good operating condition. It is also assumed that the permeability of the backfill material for the subsurface drain is at least as high as that of the surrounding soil.

REFERENCES

Allen, R. R., V. I. Myers, and L. R. Ussery. 1962. Comparison of tile and open drains for subsurface drainage on nonirrigated lands. *Am. Soc. Agr. Eng.*, Paper 62, p. 727.

Iowa Drainage Guide. 1955. *Iowa Agricultural Experiment Station. Special Report 13.*

Power Cranes-Shovels-Draglines. Technical Bull. 4. Power Crane and Shovel Association, Chicago, Illinois.

Weir, W. W. 1929. Drainage in the Sacramento Valley Rice Fields. *Univ. of Calif. Agric. Exp. Sta. Bull.* 464.

Chapter 13 DRAINAGE WELLS

The use of wells for the purpose of draining land is well established especially in irrigated areas. A number of areas have successfully controlled the water table by means of pumping from wells. However, there are only certain conditions under which well drainage can be expected to be successful. The soil permeability plays an important role in determining the feasibility of well drainage. Economics also plays a significant part.

There are two classes of wells used for drainage purposes. The water-table or gravity well is located in an unconfined aquifer and removes water directly from the root zone of the plant. It may be either a shallow or a deep well.

The other class of well encountered in drainage work is the artesian well that taps an aquifer containing water under pressure. The aquifer is confined by less permeable layers lying above it. Wells that tap artesian aquifers may be pumped or they may be free flowing. In either case the basic theory underlying their operation is the same.

Factors Affecting the Feasibility of Drainage Wells

SOIL CONDITIONS. The permeability of the soil layers which are to be drained are of paramount importance in the success of the use of drainage wells. The drainage water must be able to percolate down to the aquifer being pumped. The water table must be continuous from the shallow soil layers down to the pumped aquifer, and the permeability of the intervening layers must not present a barrier to the water movement. Some of the most successful drainage wells have been located in deep sandy soils having a relatively high permeability. Others have been located in fractured lava rock that underlies a permeable soil. In both cases the surface water table is responsive to the pumping that takes place from the aquifer. In some instances it has been possible to use pump drainage in an aquifer of limited extent in order to drain rather small areas, however the success is greater when the aquifer is of great extent.

Before locating wells in the area to be drained it is essential that a thorough investigation be made of the soil conditions. It is desirable to determine as far as possible the thickness and area of the aquifer from which pumping will take place. Test holes for the purpose of examining the soil and of

Figure 13-1 A cable tool rig for drilling a well. The bailer in the picture is used to remove the material from the well.

conducting pumping tests should be drilled in enough locations to give a reasonable definition of the feasibility of pump drainage.

AQUIFER CONDITIONS. The aquifer should be of sufficient area to insure that adequate drainage will be achieved. Most of the area to be drained should be underlain by the aquifer. The characteristics of the aquifer should be such that a large quantity of water can be pumped from it.

It is not entirely clear how deep the aquifer should be beneath the ground surface, but most of the trials of pumping from shallow aquifers, 10 feet or less beneath the soil surface, have had limited success. Many successful drainage wells are pumping from aquifers that are 80 to 300 feet beneath the ground surface. The thickness of the aquifer in combination with the permeability of the aquifer determines the rate of flow towards the well for any specified gradient.

Figure 13-2 A well used to relieve artesian pressure in a shallow aquifer about 20 feet below the soil surface.

Figure 13-3 Test pumping a drainage well to determine the aquifer characteristics.

SPACING OF WELLS. A pumping test from a trial well can be used to estimate the spacing that is necessary. Observation wells should be spaced radially from the well in four directions. The spacing of the observation wells should give an accurate picture of the cone of influence of the well. The cone of influence is the funnel-shaped drawdown of the water table around a pumped well. The curve of the water table close to the well is very steep. In order to get accurate values for the shape there should be a measurement made of the water level in the well. A piezometer should be located within about 10 feet of the well. Additional piezometers may be spaced 20, 40, 100 or more feet from the well. The drawdown of the water table caused by pumping should be observed. The results are then plotted on a map of the area. It is assumed that the effects of adjacent wells are additive. That is, if the drawdown at a location 1,000 feet from one well is 2 feet, and if another well is located 1,000 feet from the same spot in another direction, the total drawdown will be 4 feet because of the two wells.

It is possible to use the procedure described above to locate the best spacing of the wells for the most economical operation of the project. However the soils in the area must be uniform if the method outlined above is to work. The data obtained in the pumping test is used as a guide for the

Figure 13-4 Drainage well in the Turlock Irrigation District, California. There are more than 180 of these wells pumping from aquifers up to 300 feet below the soil surface. The drainage water is pumped into the irrigation canal and the water is used for irrigation purposes. Wells are spaced about one per square mile.

location of additional wells. As additional wells are dug and additional data obtained from pumping tests the criteria is modified to meet the changing soils conditions that are usually encountered.

Design of Drainage-Well Systems

GRAVITY WELLS IN UNCONFINED AQUIFERS WITH HORIZONTAL REPLENISHMENT. Gravity wells in unconfined aquifers do not have an impermeable layer lying above the aquifer. There are several different approaches to the development of equations for the operation and performance of such wells.

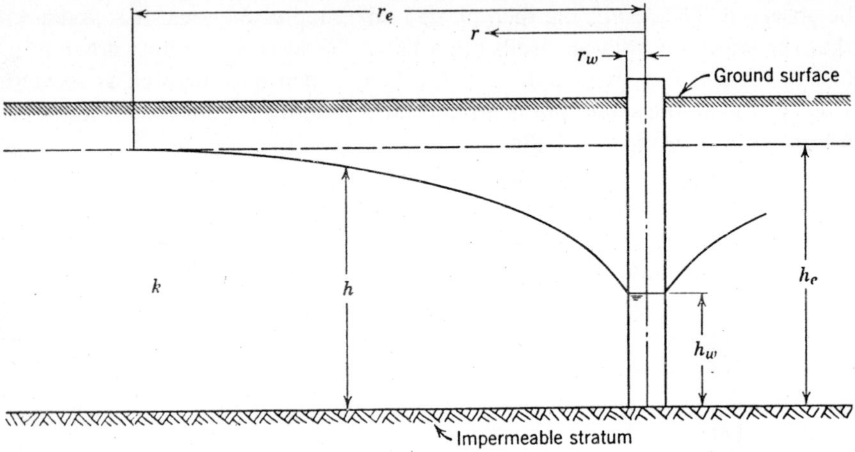

Figure 13-5 A gravity well having water-table shape according to Dupuit assumptions.

The oldest approach is that of Dupuit. The seepage surface around the well is ignored. It is assumed that the water table intersects the water level in the well. Dupuit also assumed that the flow towards the well followed horizontal paths.

For steady flow the entire discharge of the well must pass through a concentric cylinder at any radius. If we let h be the height of the water table above the plane through the bottom of the well, the hydraulic gradient will be given by dh/dr, which is the slope of the water table at radius r from the well.

The total flow through a concentric cylinder will be given by Darcy's law as

$$Q = 2\pi rhk \frac{dh}{dr} \tag{1}$$

On integration we have

$$\frac{Q}{\pi k} \ln r = h^2 + C \tag{2}$$

The head in the well is assumed to be constant with a value of h_w so that over the surface of the well, r_w, the head is equal to h_w. If this boundary condition is substituted into equation 2 the result is

$$\frac{Q}{\pi k} \ln \frac{r}{r_w} = h^2 - h_w^2 \tag{3}$$

It is now assumed that the hydraulic head remains constant at some distance r_e from the well. At r_e the head is taken to be equal to h_e. On substitution we have

$$\frac{Q}{\pi k} \ln \frac{r_e}{r_w} = h_e^2 - h_w^2 \tag{4}$$

The above equation has been shown to be quite accurate for calculating the flow into a well in cases where the conditions approximate those set down in the assumptions.

If we combine equations 3 and 4 we have an equation for the water table, in this case referred to as the free water surface.

$$h^2 = \frac{h_e^2 - h_w^2}{\ln (r_e/r_w)}\left(\ln \frac{r}{r_w}\right) + h_w^2 \tag{5}$$

This is called the Dupuit solution for the free surface. It gives quite accurate values for the free surface at large distances from the well. The agreement with measurements close to the well, however, is poor. The deviation is due to the fact that the Dupuit formula does not take into account the surface of seepage at the well. This surface of seepage is not due to the hydraulic head loss at the well, as was commonly believed some years ago. It is caused rather by the convergence of the streamlines as they near the well. Surface of seepages occur also in the flow into an open ditch or flow through an earth dam. In the use of the Dupuit formula r_e is called the radius of influence. It is arbitrarily determined and must be initially assumed.

UNCONFINED FLOW REPLENISHED BY VERTICAL PERCOLATION AS WELL AS HORIZONTAL FLOW. In irrigated areas the replenishment of the well usually results from vertical percolation of water down to the aquifer being pumped. The case of horizontal replenishment treated by the Dupuit formula is limited to those areas where vertical replenishment is not significant.

Formulas can be derived for vertical replenishment which use much the same approach used in the derivation of the Dupuit formula. In addition to the horizontal replenishment it is assumed that water is percolating into the influence region of the well at a constant rate, v. The replenishment water may come from rain falling on the soil surface. It may be from steady seepage upward from a deeper artesian aquifer, or it may be from irrigation water applied to the soil surface. In any case it is assumed to be steady and unvarying.

In deriving the equation we proceed in a manner similar to that used in deriving the Hooghoudt equation for the elliptical water table in equilibrium with the rainfall. The horizontal flow through a vertical section of the aquifer is analyzed. It is assumed that the entire flow is in a horizontal direction. The flow can be divided into two parts: one caused by horizontal replenishment and one by vertical replenishment. The vertical replenishment results from the rate of replenishment, v, acting over a surface πr^2 where r is the radial distance from the well.

The flow through the section will be given by

$$Q_r = Q - v\pi(r^2 - r_w^2) \tag{6}$$

where Q is the total flow into the well and r_w is the radius of the well. Normally r_w is small compared to r and can be neglected in the treatment that follows.

Therefore

$$Q_r = Q - \pi v r^2 \tag{7}$$

By Darcy's law we have another equation for the flow through the section. If we let the slope of the water table above the section dh/dr equal the hydraulic gradient (Dupuit-Forchheimer assumptions), we can write the following expression for the flow through the section

$$Q_r = 2\pi r h k \frac{dh}{dr} \tag{8}$$

If we equate 8 to 7 we have

$$Q - \pi v r^2 = 2\pi r h k \frac{dh}{dr} \tag{9}$$

On integration

$$\frac{Q}{\pi k} \ln r - \frac{vr^2}{2k} = h^2 + C \tag{10}$$

On the surface of the well $r = r_w$ and the head is equal to h_w. Substitution of these values in 10 gives

$$\frac{Q}{\pi k} \ln \frac{r}{r_w} - \frac{v(r^2 - r_w^2)}{2k} = h^2 - h_w^2 \tag{11}$$

At the radius of influence r_e, the head is equal to h_e. Substitution in 11 gives us

$$\frac{Q}{\pi k} \ln \frac{r_e}{r_w} - \frac{v(r_e^2 - r_w^2)}{2k} = h_e^2 - h_w^2 \tag{12}$$

Since r_e is very much greater than r_w, we can introduce the approximation

$$r_e^2 - r_w^2 = r_e^2 \tag{13}$$

If we define n as the ratio of the discharge of water resulting from vertical percolation to the total discharge of the well, then

$$n = \frac{\pi r_e^2 v}{Q} \tag{14}$$

and equation 12 becomes

$$Q = \frac{\pi k(h_e^2 - h_w^2)}{2.303 \log_{10} r_e/r_w - n/2} \tag{15}$$

The rate of replenishment is given by

$$v = \frac{nk(h_e^2 - h_w^2)/r_e^2}{2.303 \log_{10} r_e/r_w - n/2} \tag{16}$$

From equations 11 and 16—if we ignore the term (r_w^2/r_e^2)—the equation for the drawdown curve is

$$h^2 = \frac{(h_e^2 - h_w^2)(2.303 \log_{10} r/r_w - n/2(r/r_e)^2)}{2.303 \log_{10} r_e/r_w - n/2} + h_w^2 \tag{17}$$

ARTESIAN WELLS WITH CONFINED FLOW. Flow towards an artesian well that penetrates a confined aquifer can be analyzed in a manner similar to that used for unconfined flow. The flow in a confined aquifer is truly horizontal. The flow is caused by the pressure gradient in the aquifer. The pressure gradient is expressed as the change in piezometric head with the change in the radial distance. The flow is accurately expressed by

$$Q = 2\pi rmk \frac{dh}{dr} \tag{18}$$

where dh/dr is the slope of the piezometric head
m is the thickness of the aquifer
r is the radial distance from the well

Integrating equation 18 yields

$$Q = 2\pi km \frac{h_2 - h_1}{\ln r_2/r_1} \tag{19}$$

If the well completely penetrates the aquifer there is no vertical component of flow. For artesian wells there is no seepage surface at the well. If r_2 and r_1 are the radius of influence and the radius of the well respectively, $h_2 - h_1$ equals the drawdown D at the well and the equation may be written

$$Q = \frac{2\pi kmD}{\ln r_e/r_w} \tag{20}$$

In case the well only partially penetrates the aquifer, special solutions have been developed. However if there is a 50% penetration the equations given above are quite accurate.

TRANSIENT-STATE FORMULAS. The analysis of the transient conditions of well drainage has received a great deal of attention in recent years. Although the steady-state analyses presented earlier are very useful, most of the drainage situations are essentially transient. However, if the changes are taking place slowly, the steady-state formulas can be used with a high degree of precision. It is only for rapidly moving water tables that an analysis of the transient or unsteady-state condition of flow is required.

Transient conditions prevail if there is no replenishment, or if the rate of replenishment is less than the discharge of the well. The water removed from the soil by the well comes out of storage. This water is released from the soil in the same manner as water is released under the falling water-table conditions described in the section on drainage design.

Boussinesq's equation can be used to analyze the transient flow systems. Since the flow towards a well is radial in character it is necessary to write the equation in cylindrical coordinates. When the equation is expressed in cylindrical coordinates there is a reduction in the number of variables to be considered, because there is no flow around the well. The angle coordinate drops out of the equation. The differential equation is

$$\frac{S}{km}\frac{\partial h}{\partial t} = \frac{\partial^2 h}{\partial r^2} + \frac{1}{r}\frac{\partial h}{\partial r} \tag{21}$$

where S is the storage coefficient or specific yield. It is generally considered equal to the volume of water released from a unit area of soil for a unit drop in the piezometric head. As used in the equation the storage coefficient is considered a constant.

A more precise representation of the storage coefficient is given in the discussion of the drainable pore volume in Chapter 6. For many practical cases the drainable pore volume can be represented by a single number that is equivalent to the storage coefficient. However, care should be used in applying the concept to the case of a rising water table. In the case of rising water table the soil moisture content at the beginning of the rise (called the antecedent soil moisture) must be considered in evaluating the storage coefficient.

In equation 21 km may be represented by T, which is called the transmissivity of the aquifer.

In the case of artesian wells it is considered that the water given by the storage coefficient is water that is released because of consolidation and compression effects associated with the reduction of pressure in the aquifer. *Nonsteady-State Artesian Wells in an Artesian Aquifer.* The following expression has been derived by Theis (1935) for a constantly discharging well that drains an artesian aquifer of infinite extent. For this case the drawdown, $D = h_0 - h$, at any radial distance r and time of pumping t, is given by

$$D = \frac{Q}{4\pi T}\int_{r^2S/4Tt}^{\infty} \frac{e^{-u}}{u}\, du \tag{22}$$

where S is the specific yield. The integral on the right side of the equation is known as the exponential integral and is extensively tabulated. In the literature on wells the integral is often written as

$$W(u) = \int_{u}^{\infty} \frac{e^{-u}}{u}\, du \tag{23}$$

where $W(u)$ is termed the well function. The above equations are now rewritten

$$u = \frac{r^2 S}{4Tt} \tag{24}$$

and

$$D = \frac{Q}{4\pi T} W(u) \tag{25}$$

If it is desired to find values for S and T from pump-test data, it is necessary to use a graphical method utilizing Figure 13-6. This figure shows $W(u)$ plotted as a function of u. From equations 24 and 25 for a particular well test, u is proportional to r^2/t and D is proportional to $W(u)$. By plotting D as the ordinate and r^2/t as the abscissa on transparent paper to the same scale as Figure 13-6, a curve will be obtained that is similar to the one in

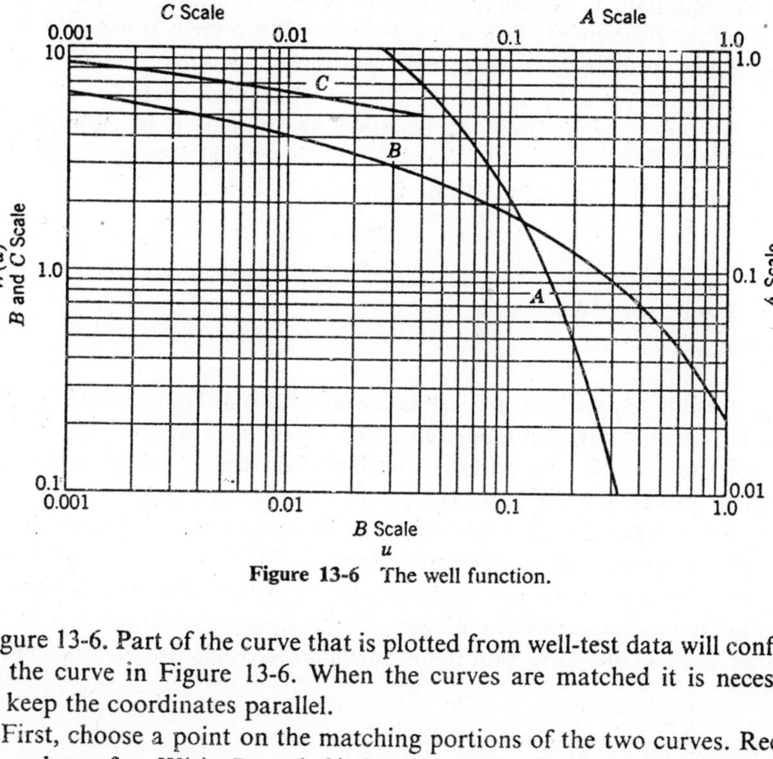

Figure 13-6 The well function.

Figure 13-6. Part of the curve that is plotted from well-test data will conform to the curve in Figure 13-6. When the curves are matched it is necessary to keep the coordinates parallel.

First, choose a point on the matching portions of the two curves. Record the values of u, $W(u)$, D, and r^2/t for this point. The values of D and $W(u)$ may then be substituted into equation 25 to solve for T. Using this value of T, substitute T, u, and r^2/t into equation 24 to solve for S. For drawdown measured at the well, r may be taken as the well radius r_w.

For most practical cases an approximation of equation 25 will be sufficiently accurate. The exponential integral may be expanded into an infinite series to give

$$D = \frac{Q}{4\pi T}\left[-0.5772 - \ln\frac{r^2S}{4Tt} + \frac{r^2S}{4Tt} - \frac{1}{2.2!}\left(\frac{r^2S}{4Tt}\right)^2 + \frac{1}{3.3!}\left(\frac{r^2S}{4Tt}\right)^3 - \cdots\right] \tag{26}$$

For relatively small values of u (small values of r or large values of t), all except the first two terms of equation 26 are insignificant and

$$D = \frac{Q}{4\pi T}\left(\ln\frac{1}{u} - 0.5772\right) \tag{27}$$

Substitution of $u_1 = r^2S/4Tt_1$ and $u_2 = r^2S/4Tt_2$ successively in equation 26 and subtraction gives, for a particular radial distance,

$$D_2 - D_1 = \frac{0.183Q}{T}\log_{10} t_2/t_1 \tag{28}$$

To use equation 28, D is plotted as a function of time on semi-logarithmic paper. Initially the points follow a curved line but after a short period the line becomes straight. It is the straight portion of the curve that is used to solve for T. If D_1 and D_2 are chosen for one logarithmic cycle, $\log t_2/t_1 = 1$, which facilitates the solution of equation 28. If $D = 0$ in equation 27, then $\ln(1/u_0) = 0.5772$ and $u_0 = 0.563$. If the straight line is extended to $D = 0$ to obtain t_0, substitution into equation 24 yields

$$S = \frac{2.25Tt_0}{r^2} \tag{29}$$

which may be used to solve for S.

Gravity Wells

It is possible to apply the artesian nonsteady formula to the case of gravity wells. However, a correction must be made for the area of flow. The area of flow for the artesian well remains constant. The area of flow is constantly changing for the gravity well because of the recession of the water table. Jacobs has proposed that a correction be applied for the changing area of flow. He indicates that a correction equal to $D^2/2m$ should be subtracted from the observed values of the drawdown D. In this case m is taken to be equal to the original depth of the water. The analysis then follows that for the artesian well. If the correction is not made the answers are in considerable error.

The Determination of the Aquifer Characteristics

The properties of an aquifer can be represented by the following data.

1. The thickness
2. The hydraulic conductivity or permeability
3. The specific yield, or storage coefficient

The transmissivity of an aquifer is the product of the thickness of the aquifer and the hydraulic conductivity. It is the volume of water that will flow in a unit time under a unit hydraulic gradient through a vertical strip of a water-bearing material of unit width, extending the full saturated thickness of the formation.

THE THEIS METHOD FOR UNCONFINED WELLS. Observation wells are located at distances r_1 and r_2 from the center line of the well. The well is placed in operation and is pumped until equilibrium conditions are reached. When there is no longer any change in the water levels in the well, the amount of drawdown in the observation wells is measured. The drawdown in the observation wells is D_1 and D_2. The transmissivity is then calculated using the equation

$$T = \frac{527.7Q \log_{10}(r_2/r_1)}{D_1 - D_2} \tag{30}$$

where Q is the volume per unit time of discharge from the well, and 527.7 is a conversion factor to convert the measurement to gallons per foot per day. All measurements in the above equation are in feet and in gallons per minute.

The Theis equation assumes that equilibrium conditions prevail. It is not often that the conditions in the field represent a true equilibrium.

THE THEIS METHOD FOR THE UNSTEADY-STATE CONDITIONS OF AN UNCONFINED WELL. Theis has presented an analysis that is based on the decline rate of the water level in a single piezometer within the cone of pressure relief. The derivation, which is not presented here, is based on the premise that all the water discharged from the pumped well is derived from storage in an extensive aquifer.

Theis's equation is

$$D = \frac{114.6Q}{T} \int_{1.87r^2S/Tt}^{\infty} \left(\frac{e^{-u}}{u}\right) du \tag{31}$$

where D = drawdown of piezometric level in feet
Q = discharge of pumped well in gallons per minute
r = distance of piezometer from pumped wall in feet
T = transmissivity coefficient in gallons per day per foot
S = coefficient of storage as a ratio or decimal
t = time that the well has been pumped in days

The solution of this equation is simplified by the use of the well function, $W(u)$ as described above.

The exponential function can be rewritten as

$$D = \frac{114.6Q}{T} W(u) \tag{32}$$

The value of the integral can be found by the solution of a series where

$$u = \frac{1.87r^2S}{Tt}. \tag{33}$$

A graphical method has been devised by Theis, and its use greatly simplifies the use of the nonsteady formula. Values of $W(u)$ are plotted on logarithmic paper as a function of u. This curve is known as a "type curve." The previous equations are rewritten as

$$D = \frac{114.6Q}{T} = W(u) \tag{34}$$

and

$$\frac{r^2}{t} = \frac{T}{1.87S}u \tag{35}$$

For any given pumping test the values within the brackets should be a constant, provided the assumptions made in the derivation are met. Since D is related to r^2/t as $W(u)$ is related to u, the drawdown, D, occurring in any piezometer can be plotted against the values of r^2/t on logarithmic tracing paper to the same scale as the type curve $W(u)$ versus u. Then, by superposition, these curves can be aligned and a matching point selected. Values for the calculation of T and S can then be taken from the curve.

REFERENCES

Babbitt, Harold E., and David H. Caldwell. 1948. The free surface around and interference between gravity wells. *Illinois Engineering Experiment Station Bull.* 374.

Engelund, Frank. 1957. On the theory of multiple-well systems. *Acta Polytechnica, Civil Engineering and Building Construction Series*, Vol. 4, No. 7, Copenhagen, Denmark.

Hantush, M. S. 1957. Nonsteady flow to flowing wells in leaky aquifers. *J. Geophys. Res.*, **64**:1043–1052.

Jacob, C. E. 1946. Radial flow in a leaky artesian aquifer. *Trans. Am. Geophys. Union*, **27**:198–205.

Marr, James C. 1926. Drainage by means of pumping from wells in Salt River Valley, Arizona. *U.S. Dept. Agr. Bull.* 1456.

Muskat, M. 1946. *The Flow of Homogeneous Fluids Through Porous Media.* McGraw-Hill Book Co. (reprinted by Edwards Bros., Ann Arbor, Michigan).

Peterson, Dean F., Jr., O. W. Israelsen, and V. E. Hansen. 1952. Hydraulics of Wells. *Agricultural Experiment Station Bull.* 351 (*Technical*) Utah State Agricultural College.

Scott, V. H., and J. N. Luthin. 1959. Investigation of an artesian well adjacent to a river. *Proc. Am. Soc. Civil Eng., Irrigation and Drainage Division* 85, No. IR 1.

Theis, C. V. 1935. The relation between the lowering of the piezometric surface and the rate and duration of discharge of a well using groundwater storage. *Am. Geophys. Union Trans.*, **16**:519–524.

Wenzel, L. K. 1942 Methods for determining the permeability of water bearing materials. *U.S. Geol. Survey, Water Supply Paper* 887.

PROBLEMS

1. Use the Dupuit formula for a gravity well in unconfined aquifers with horizontal replenishment for the following problem. The drawdown in the well is 30 feet. The water table is 3 feet below the soil surface, at a distance r_e of 1500 feet. The well is 24 inches in diameter.

2. How far apart must wells be spaced in order that the water table be more than 6 feet from the soil surface?
3. In the above problem water is added to the soil surface at the rate of 1.5 inches per hour. All of this water percolates down to the water table. Calculate the drawdown curve for one well.
4. An artesian well completely penetrates an aquifer that is 25 feet thick. The drawdown at the well is 40 feet. The well discharge is 2000 gpm. The well radius is 24 inches and the radius of influence is 1500 feet. What is the hydraulic conductivity of the aquifer? What is the transmissivity of the aquifer?

Chapter 14 SURFACE DRAINAGE AND TIDAL RECLAMATION

Large areas of land in the eastern half of the United States and Canada are excessively wet. The condition is caused largely by the inability of excessive rainfall to move over the ground surface to an outlet, or through the soil to a subsurface drainage system. These poor surface-drainage conditions are usually associated with soils that have low permeability. Often the soils are very shallow over a barrier such as rock, or very dense clay pan. The impermeable subsoil prevents the water from moving downward and prevents the proper functioning of a subsurface drainage system. Many of these areas have flat topography with surface depressions which prevent the excess rainfall from moving over the ground surface. Often the land slope is not sufficient to permit the water to flow across the ground surface. In some instances the areas lack adequate drainage outlets. In order to correct this problem something must be done to eliminate the depressions and to provide sufficient slope for overland flow. In addition, it is necessary to provide channels to convey the water from the area that is affected.

The practice of surface drainage may be defined as the diversion, or orderly removal, of excess water from the surface of the land by means of improved natural or constructed channels. The channels may have to be supplemented by shaping and grading of the land surface so that the water may flow freely into the channel.

Many times a subsurface system of drain pipes is needed in conjunction with the surface drains. The effectiveness of the subsurface drains is increased by the removal of water from the soil surface. As soon as the surface water is removed, the drain pipes can act to lower the water table and to provide a satisfactory environment for the growth of plants.

TYPES OF SURFACE DRAINAGE

There are essentially five types of surface field-drainage systems that are in common use today. Combinations of two or more of these systems may be required by the circumstances encountered in the field. The five types of systems are classified as (1) the bedding system, (2) the random ditch system, (3) the interception system, (4) the diversion-ditch system, and (5) the field-ditch system. Some of the factors which influence the choice of a

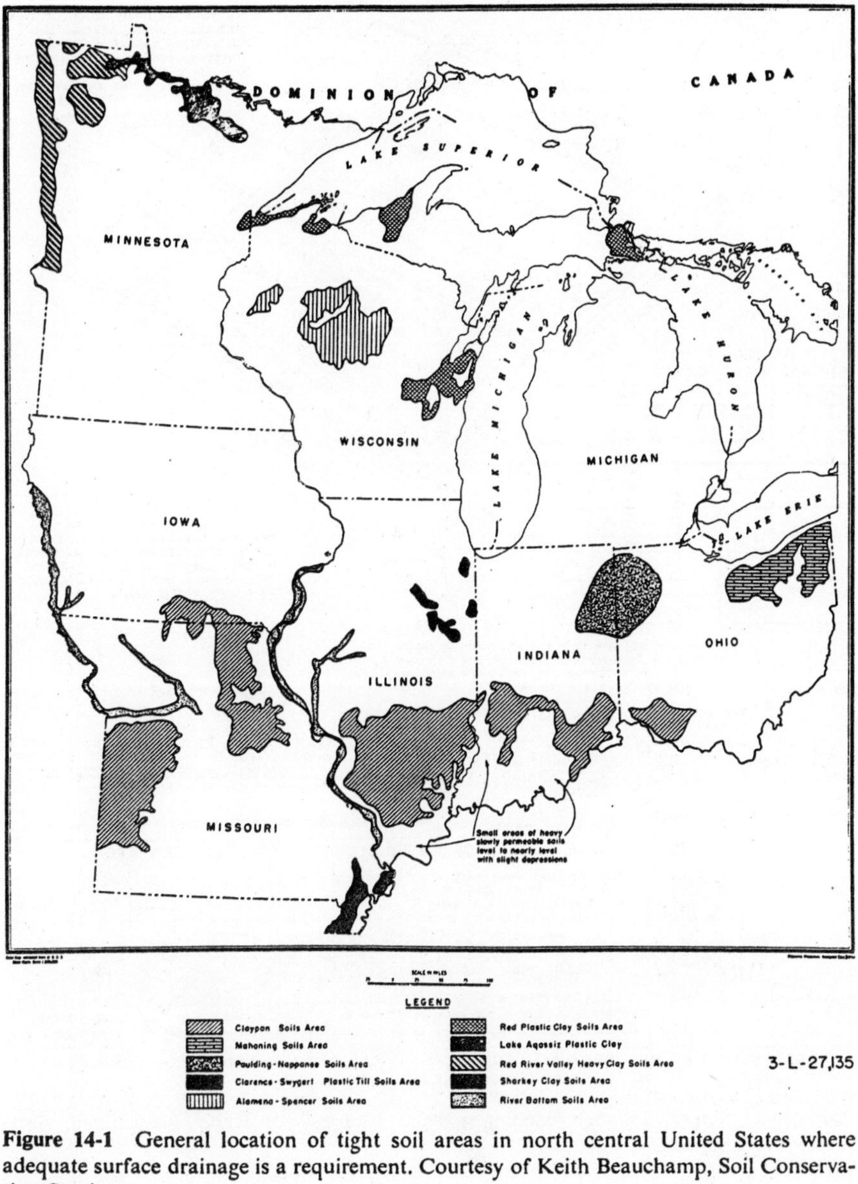

Figure 14-1 General location of tight soil areas in north central United States where adequate surface drainage is a requirement. Courtesy of Keith Beauchamp, Soil Conservation Service.

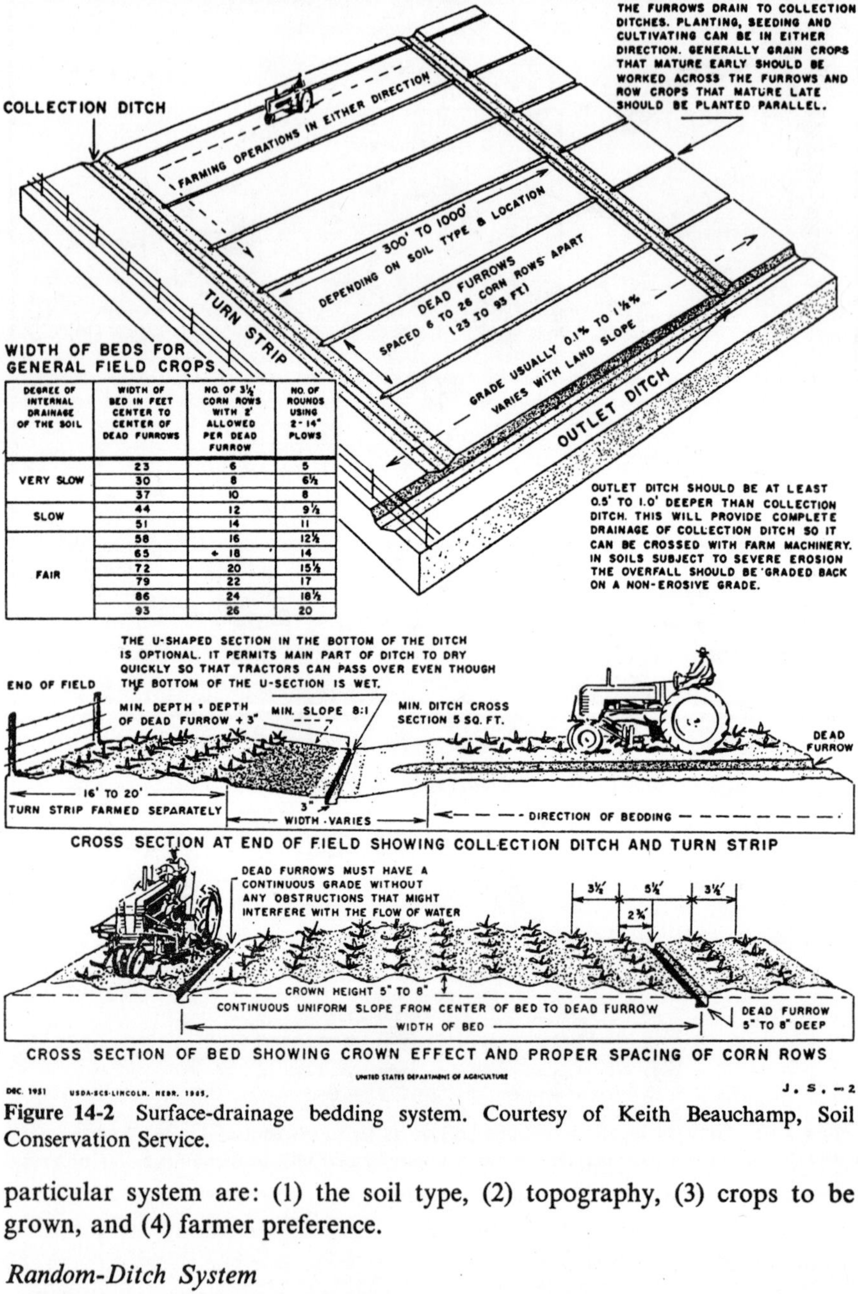

DEGREE OF INTERNAL DRAINAGE OF THE SOIL	WIDTH OF BED IN FEET CENTER TO CENTER OF DEAD FURROWS	NO. OF 3½' CORN ROWS WITH 2' ALLOWED PER DEAD FURROW	NO. OF ROUNDS USING 2-14" PLOWS
VERY SLOW	23	6	5
	30	8	6½
	37	10	8
SLOW	44	12	9½
	51	14	11
FAIR	58	16	12½
	65	18	14
	72	20	15½
	79	22	17
	86	24	18½
	93	26	20

Figure 14-2 Surface-drainage bedding system. Courtesy of Keith Beauchamp, Soil Conservation Service.

particular system are: (1) the soil type, (2) topography, (3) crops to be grown, and (4) farmer preference.

Random-Ditch System

The random-ditch system is adapted to areas that have depressions which are too deep or too large to fill by land smoothing. The surface drainage ditches may meander from one low spot to another, collecting the water and carrying it to an outlet ditch.

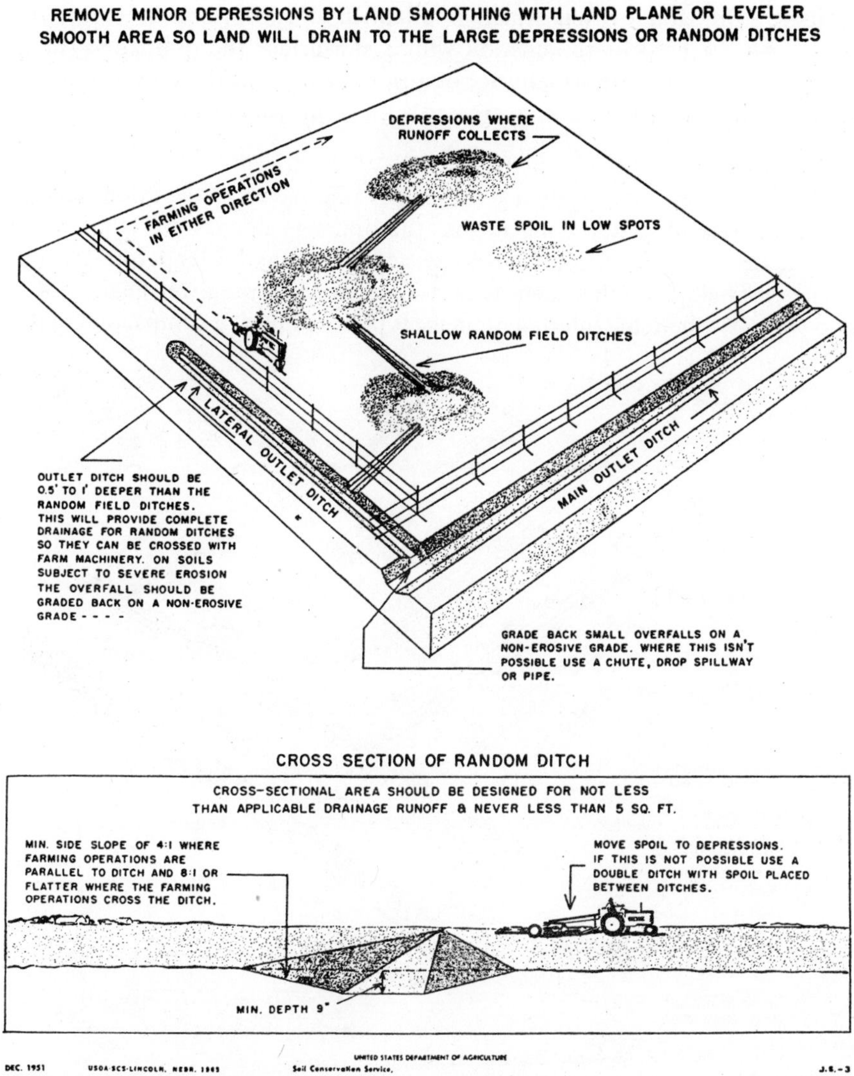

Figure 14-3 Random-ditch system of surface drainage. Courtesy of Keith Beauchamp, Soil Conservation Service.

Drainage in these areas is improved if the entire field is smoothed or graded to remove minor depression and allow the surface water to flow to the ditches.

In constructing the random ditch, the spoil from the ditches can be placed in the minor depressions that will not be drained by the ditches. The ditch must be of sufficient size and depth to drain off the impounded water rapidly and completely. On flat, very slow permeable soils, it may be necessary to combine this system with the bedding system to do an adequate job of surface drainage. On land or soil which is adapted to subsurface drainage, the system can be used in conjunction with a subsurface tile drainage system.

Since the ditches are usually located in cultivated fields, they should be constructed so that they can be crossed with farm machinery.

The Cross-Slope Ditch System or Interception System

The cross-slope ditch system is sometimes referred to as the drainage-type terrace. It resembles terracing in that the drainage ditches are constructed around the slope on a uniform grade according to the land topography.

The cross-slope ditch system is best adapted to sloping wet fields of 4% slope or less, where the internal drainage is poor because of compaction of the

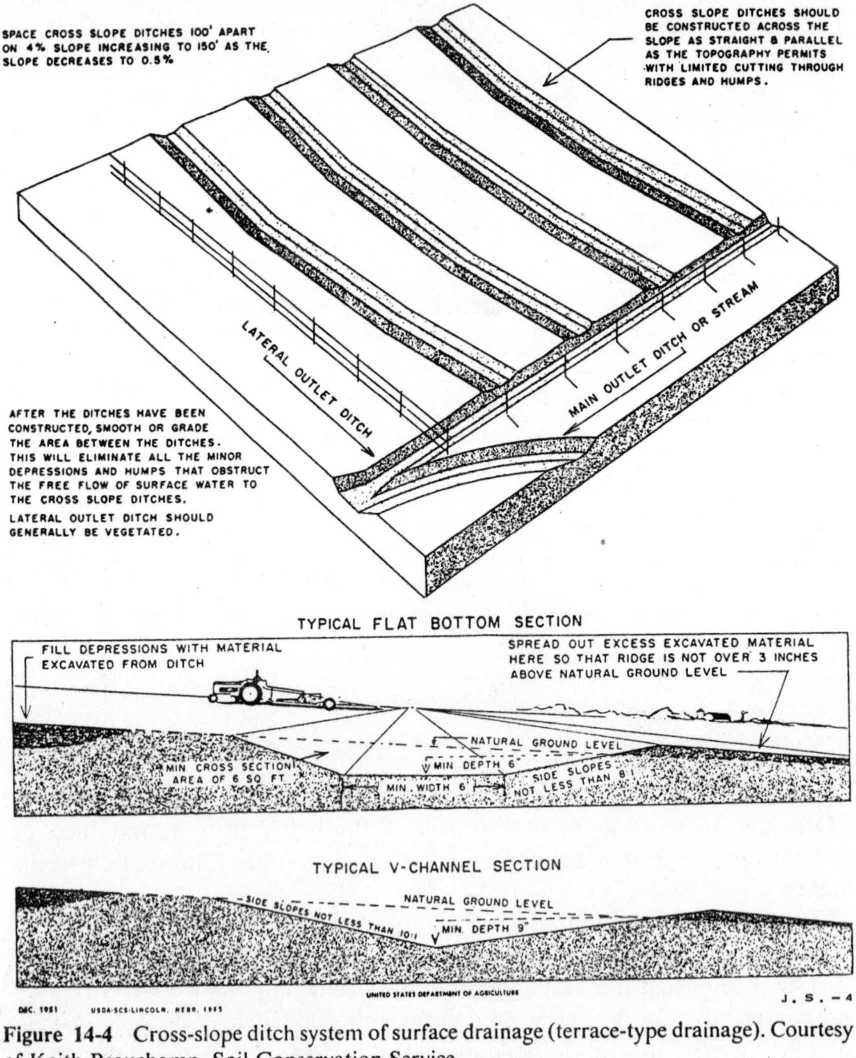

Figure 14-4 Cross-slope ditch system of surface drainage (terrace-type drainage). Courtesy of Keith Beauchamp, Soil Conservation Service.

soil, and where many shallow depressions hold the water after the rain. The soil profile is such that the water collects in the low spots after a rain and cannot infiltrate into the soil. The depressions are too numerous and the slope too great for successful bedding, and subsurface drains are generally not practical or feasible.

The cross slope ditches or terraces should be constructed across the slope as straight and parallel as topography permits. There should be limited cutting through ridges and humps. The spacing between ditches should be about 100 feet on a 4% slope and increase to 150 feet as the slope decreases to 0.5%. The ditches are built with little or no ridge on the down-slope side of the ditch. This provides for ease in crossing the ditches and reduces the damage caused by overflow. The excavated material from the ditches can be placed in the depression areas between the ditches. Any excavated material not used in this operation should be spread out on the down-hill side of the ditch. The ridge should not be over 3 inches above the natural ground surface. After the excavated material from the ditches has been placed in the depressions, the area between the ditches should be smoothed or graded to eliminate all minor depressions or humps.

The success of this system of surface drainage depends upon the elimination of the depressions between the ditches. All farming operations should be parallel to the ditches. If the ditches are laid as straight and parallel as possible, it will reduce the difficulty of farming.

Parallel- or Diversion-Ditch System

On flat, poorly drained soils that have numerous shallow depressions, the parallel-ditch system is suitable. In general, the parallel ditches are 600 to 1200 feet apart and the land between the ditches is sloped and smoothed to eliminate any minor depressions or obstructions to the overland flow of the water.

Land Smoothing

Most of the soils that require surface drainage have many depressions in the surface which vary in size and shape. They may be a foot or more deep or they may be veiy shallow. In order to keep the surface water moving at a uniform rate over the soil surface it is necessary to fill the depressions and smooth out the high points in the field. It is not always possible to ditch all the depressions. The larger depressions should be connected to a ditch, but, there are many shallow pockets which will collect and hold water for long periods after a rain. These pockets should be eliminated by land smoothing.

RECLAMATION OF TIDAL LANDS

Tidal lands are lands subject to periodic inundation due to tidal fluctuations of the sea. These lands may adjoin the open ocean or they may face an enclosed bay. The most dramatic reclamation of these lands occurs in Northern Europe in Holland, England, and parts of Germany. There are

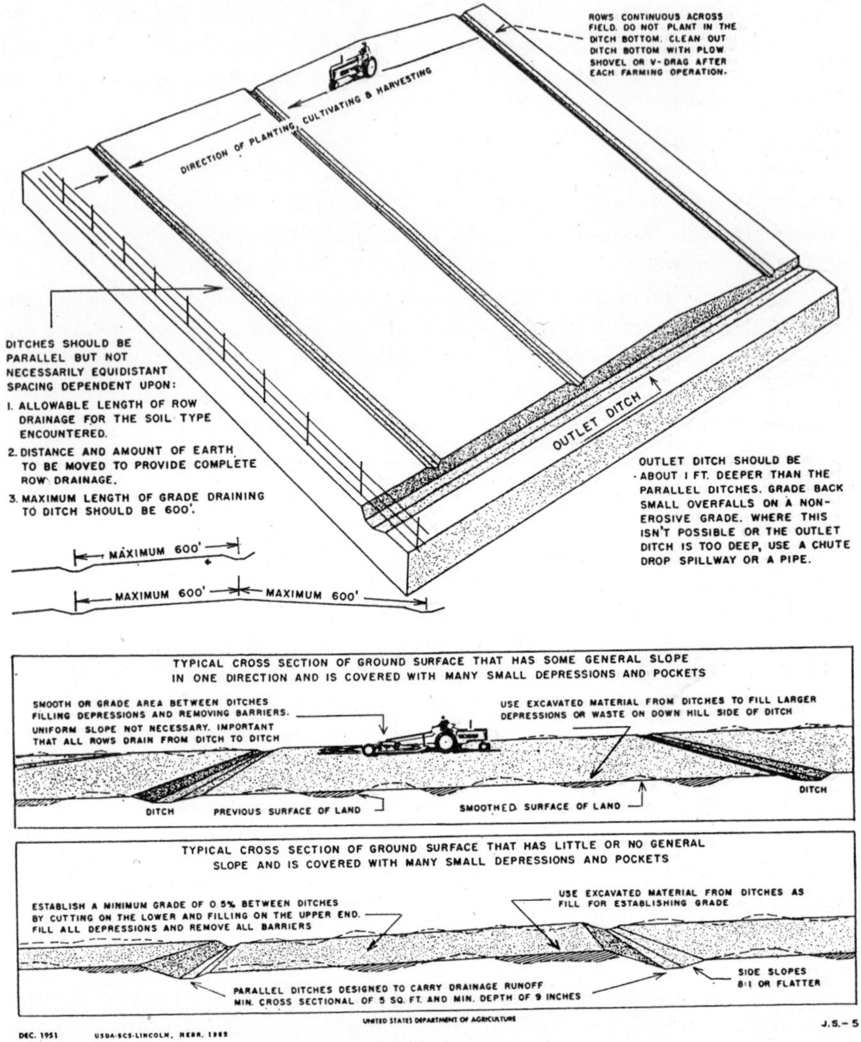

Figure 14-5 Parallel-ditch system of surface drainage. Parallel ditches to intercept and rapidly remove surface water from the field and reduce the length of row drainage. Courtesy of Keith Beauchamp, Soil Conservation Service.

extensive areas in other parts of the world, including the eastern and southern United States, that can be reclaimed from the sea.

The process of reclamation basically consists of the following steps. The erection of a dike or a levee to prevent the intrusion of the sea water, the provision of surface ditches to carry off the surface water, and the provision of a pump outlet to pump the drainage water into the ocean.

The exact steps to be followed in the reclamation of a particular area depends upon local site conditions. One of the most important factors to

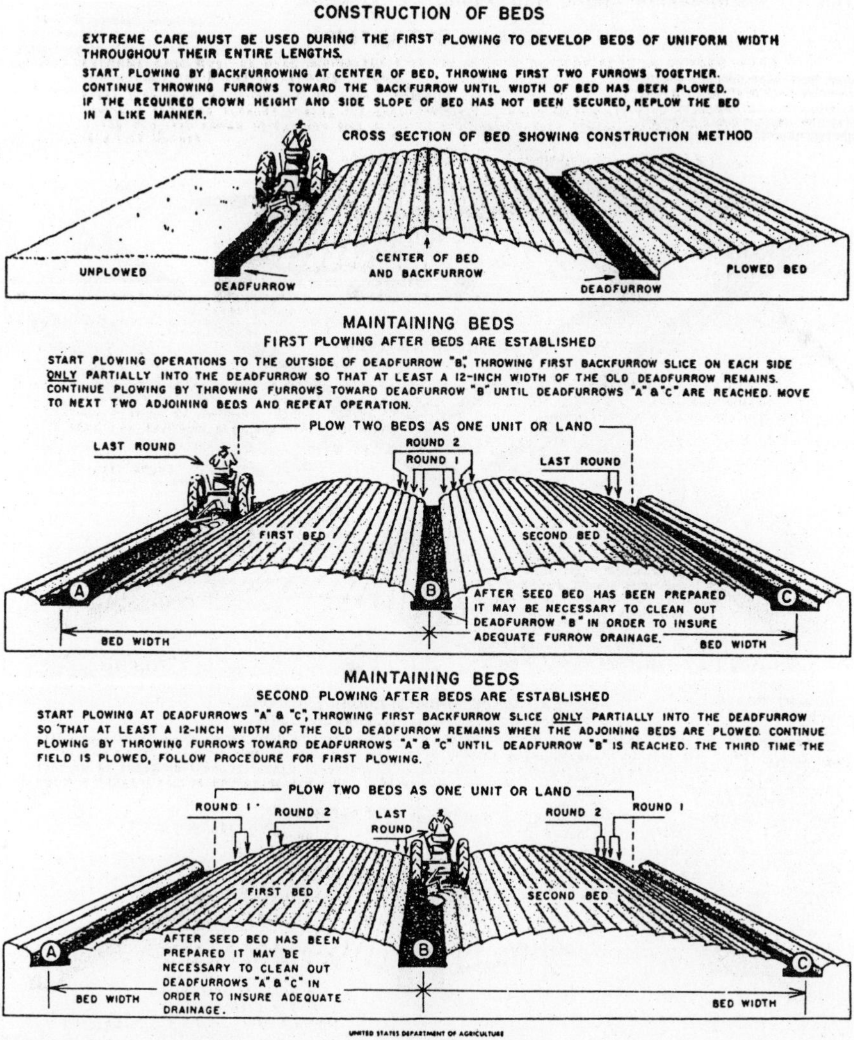

Figure 14-6 Constructing and maintaining bedding systems for surface drainage. Courtesy of Keith Beauchamp, Soil Conservation Service.

consider is the soil in the area to be reclaimed. In Holland the soil has been laid down by streams and consists of loose alluvial deposits that have very little stability. Until these soils are drained they cannot support the weight of a machine, or even, in some cases, that of a man. The drainage system is laid out prior to the pumping out of the surface water. The ditches are all dug by dredges. In most instances the use of dredges is more economical than any other method for digging ditches. After the ditches are dug, water is pumped out of the area and drainage takes place. The initial step is a consolidation of the soil and its increased stability so that it can bear the weight

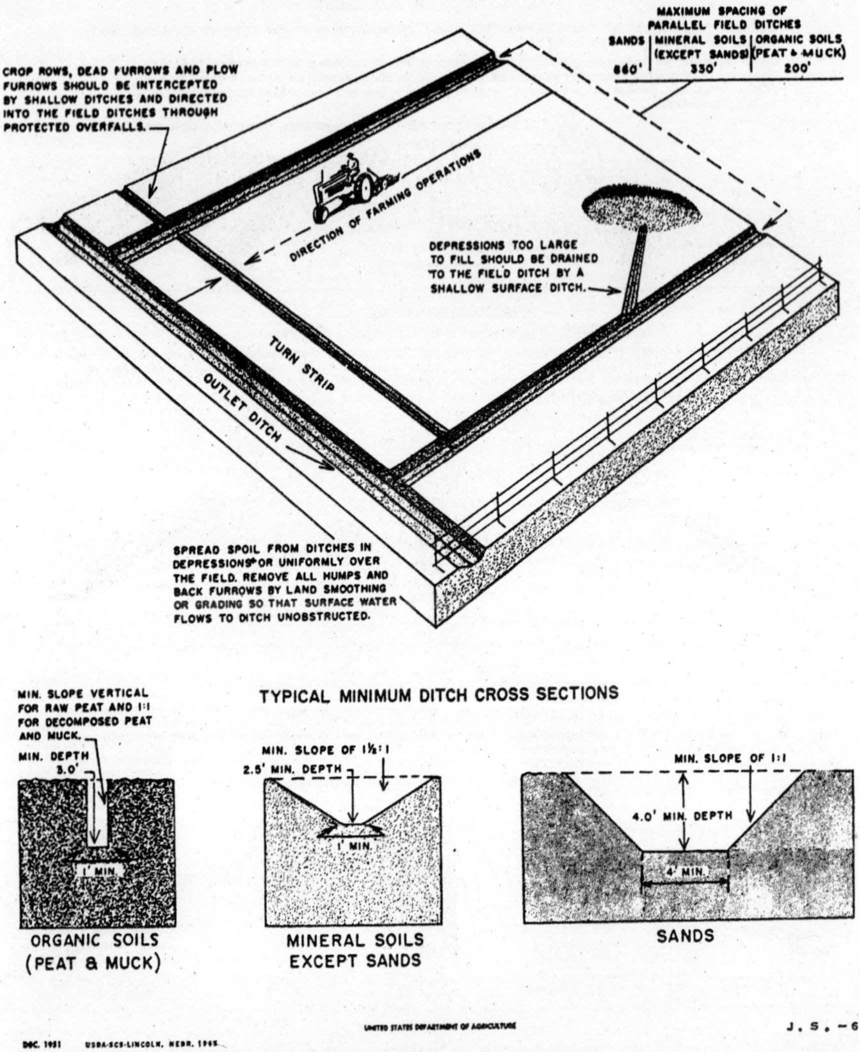

Figure 14-7 Field-ditch system for water-table control and surface water removal. Courtesy of Keith Beauchamp, Soil Conservation Service.

of machinery. When this has been accomplished, soil reclamation procedures can be instituted to remove the salt if there is such in the soil, and to start to grow crops. Normally, the first crops are salt resistant crops such as barley or rye and these can be followed with soil building crops or other crops as the conditions require.

Some of the areas to be reclaimed have soils that are very high in organic matter. These soils consist of peats and mucks that are up to 80% organic matter. Organic soils present a special problem in reclamation. These problems arise from the physical and chemical properties of organic soils.

Figure 14-8 Bedding of pasture land, Louisiana.

Figure 14-9 Land prepared for row crops, Louisiana.

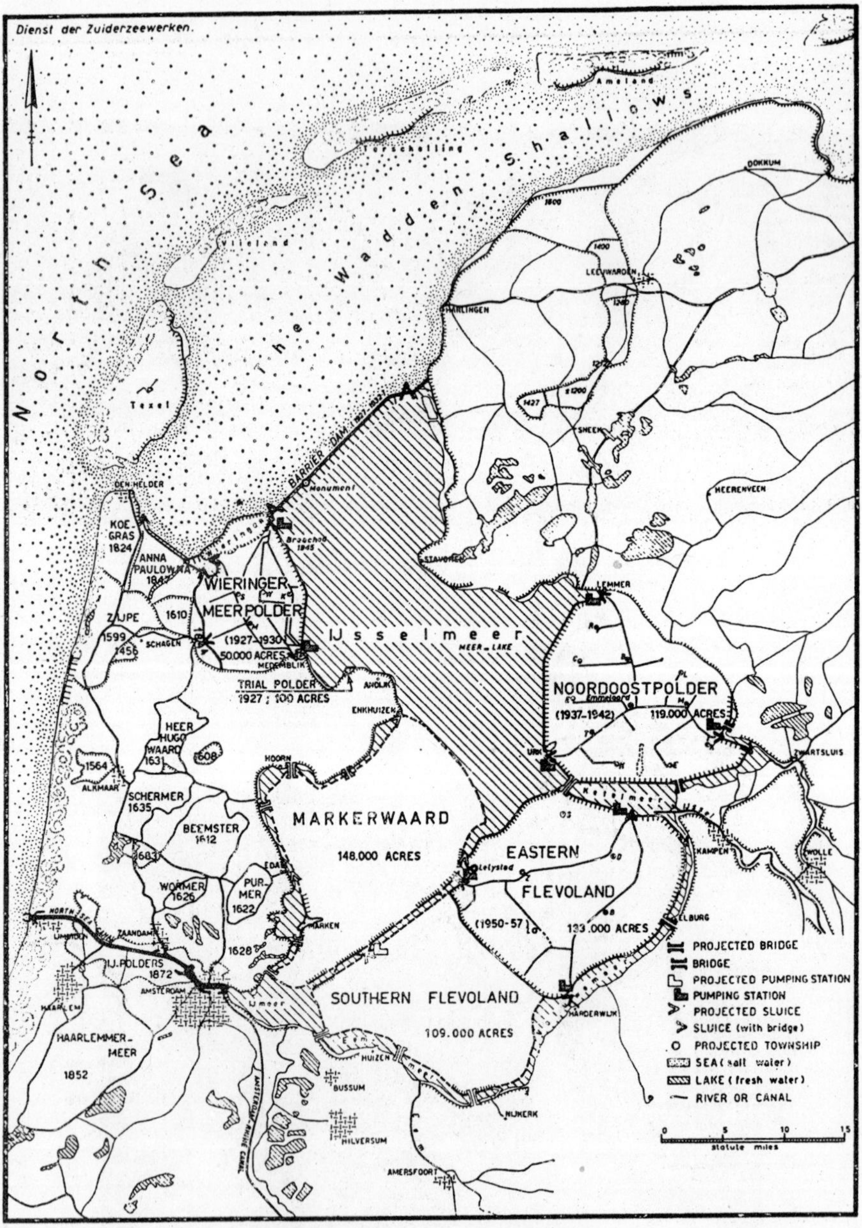

Figure 14-10 Zuiderzee polders.

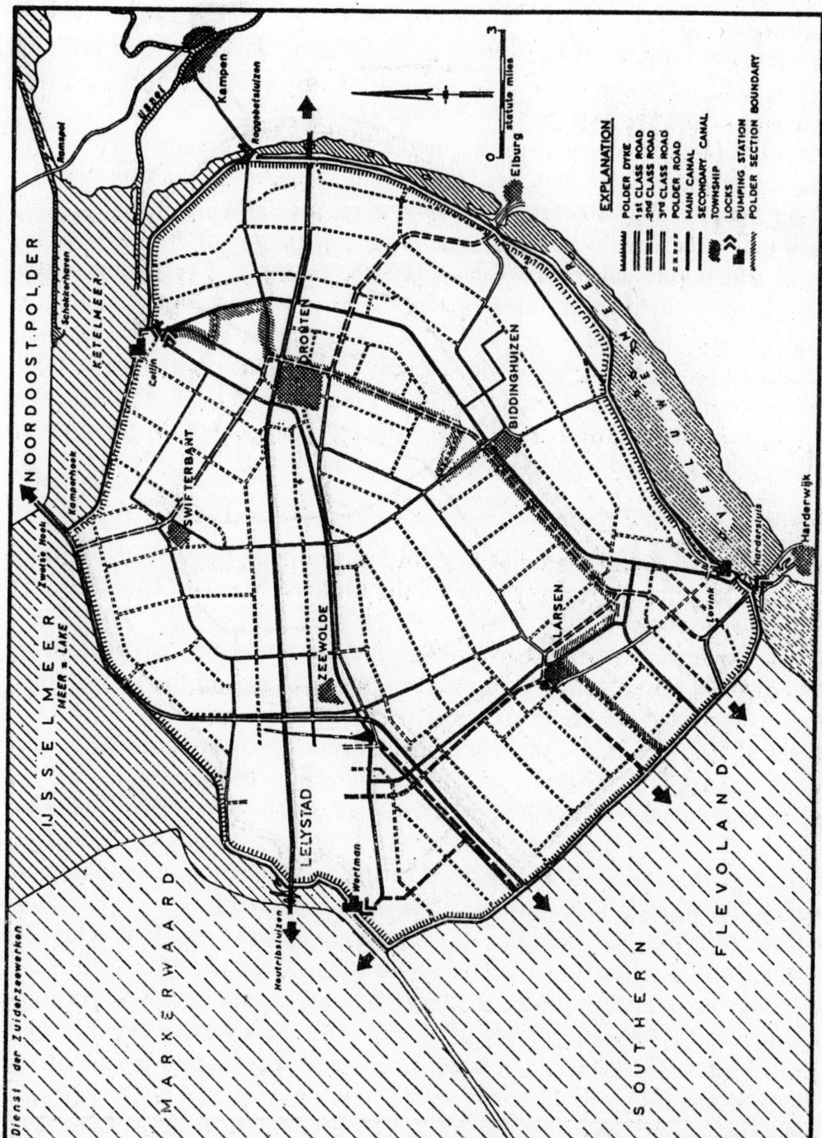

Figure 14-11 Eastern Flevoland. Provisional scheme.

Organic soils can hold large quantities of water. However, when organic soils dry there is a drastic change in their volume. This causes large cracks to form in the soil and causes excessive shrinkage of the soil mass. Sometimes the shrinkage process is not completely reversible. That is, when the soil wets up it does not swell back to exactly the same volume it had prior to shrinkage. If the soil is drained to great depths, there will be considerable subsidence of the soil surface owing to the shrinkage.

An additional factor to consider in the drainage of organic soils is the oxidation of the organic material. Some studies conducted indicate that the organic material oxidizes rather quickly when it is permitted to dry. This means that over a period of years the organic material will disappear. One way in which this can be prevented is to keep the soil relatively moist. Maintaining a high water table in these soils will reduce the amount of subsidence due to drying and will also reduce the rate at which the soil oxidizes. In most organic peat bogs there is some control of the water table to maintain this important soil resource. By maintaining the water table at a constant elevation relatively close to the soil surface much of the plant needs for water are supplied by subirrigation.

Dikes in the United States

In the United States there has been extensive diking for surface drainage in the south central area adjacent to the Gulf of Mexico. The dikes are classified as follows.

Class I dikes are designed to provide the maximum feasible protection. These dikes are used where there is a possibility of loss of life or very high values of land and land improvements.

Class II dikes or levees include embankments which are built to protect agricultural lands of medium to high productivity. In these instances, the land improvements are limited to farmstead and various farm facilities. These levees are designed to protect the land from floods of infrequent occurrence. Mineral soils can be used in their construction. The maximum design water stage is 12 feet above normal ground surface.

Class III levees are designed to protect agricultural land of relatively low value and are limited to low heads of water. They are often built from spoil from excavated drainage channels. The maximum design water stage against a class III levee is 6 feet for mineral soils and 4 feet for organic soils.

In instances where the levee faces open water, the desired height of the earth levee should be equal to the sum of the requirements of the designed depth of water and the freeboard. In the instances where the levee faces open water and wave action is anticipated, the wave height can be approximated by the formula

$$h = 1.5D^{\frac{1}{2}} + 2.5 - D^{\frac{1}{4}}$$

where h = height of wave in feet from trough to crest

D = length of exposure in miles

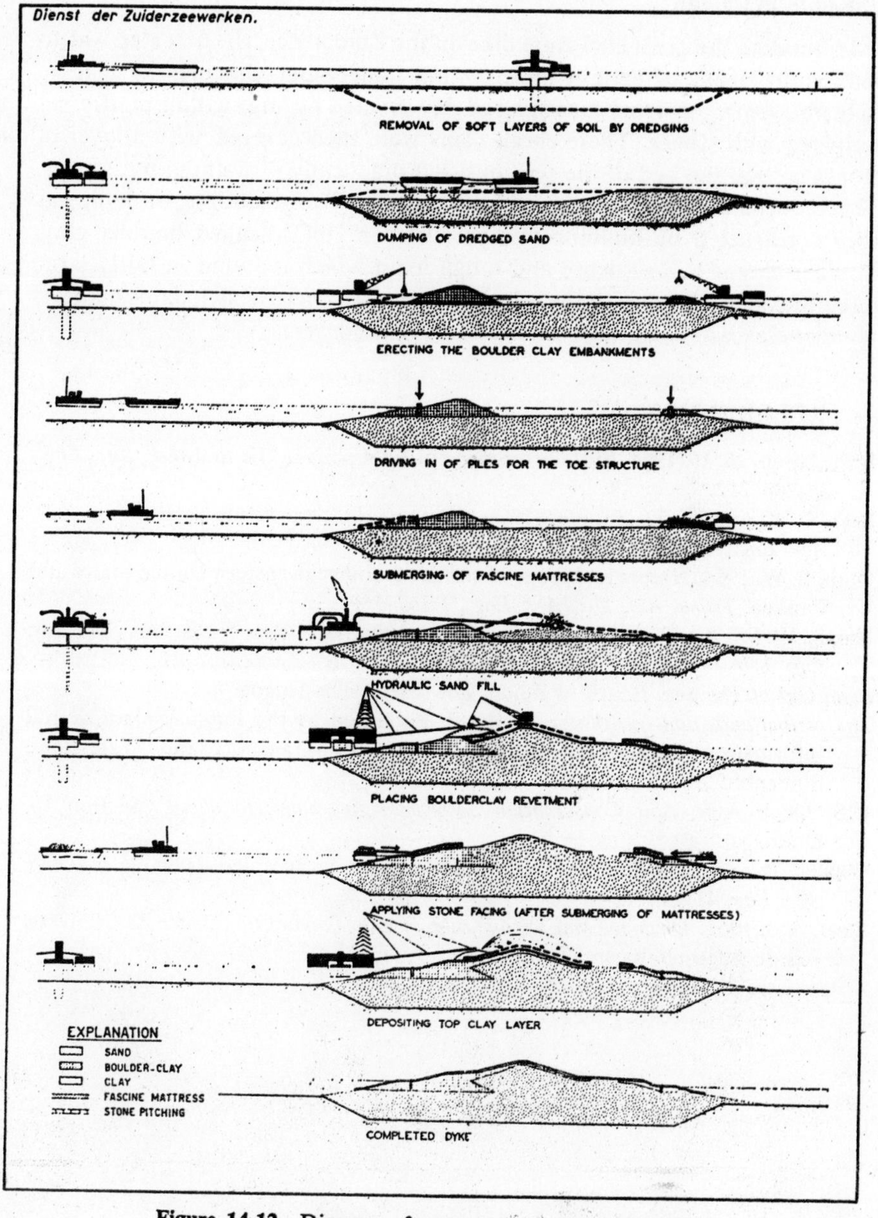

Figure 14-12 Diagram of construction of a polder dike.

Dikes in Holland

In building the great enclosure dike on the Zuider Zee, the first step was to reduce the depth of certain underground underwater channels by placing in them submerged dams consisting of clay covered by huge willow mattresses weighted with stones. These check dams were then covered with a layer of stones so that the bed of the sea consisted practically of nothing but stones. This reduced the scouring action of the currents. The great dike, or dam, was then constructed by building two parallel dams of so-called boulder clay. Boulder clay is a very dense and tough loam which is found in fairly large quantities at the bottom of the lake. The space between the two dams made of boulder clay was then filled with sand from dredgers.

REFERENCES

Beauchamp, K. H. 1952. Surface drainage of tight soils in the midwest. *Agr. Eng.*, **33**:208–212.

Beer, C. E. and W. D. Shrader. 1961. Response of corn yields to bedding soils. *Agr. Eng.*, **42**:618–621.

Gain. E. W. 1964. Nature and scope of surface drainage in eastern United States and Canada. *Trans. Am. Soc. Agr. Eng.*, **7**:167–169.

Harris, W. S. 1964. Row-crop drainage in the Arkansas Delta. *Trans. Am. Soc. Agr. Eng*. **7**:74–66.

Land Out of The Sea. Board of Zuider Zee works, The Hague.

The Netherlands and the Water. Undated. Published by the Public Relations and Information Department of the Ministry of Transport and Waterstaat, Binnenhof 20, The Hague.

U.S. Dept. Agr., *Soil Conservation Service Engineering Handbook*. Section 16, Drainage.

Walker, P. and J. H. Lillard, 1960. Land forming an accepted drainage practice. *Agr. Eng*. **41**:24–27.

Zuur, A. J. 1951. *Drainage and Reclamation of Lakes and of the Zuider Zee*. Directie van de Wieringermeer. Zwolle.

INDEX